SCHOLASTI

LITTLE LEARNER PACKETS

NUMBERS

Immacula A. Rhodes

Cover design: Tannaz Fassihi; Cover illustration: Jason Dove
Interior design: Michelle H. Kim
Interior illustration: Doug Jones

ISBN: 978-1-338-22829-8

Printed in the U.S.A.
First printing, January 2018.

1 2 3 4 5 6 7 8 9 10 40 24 23 22 21 20 19 18

Table of Contents

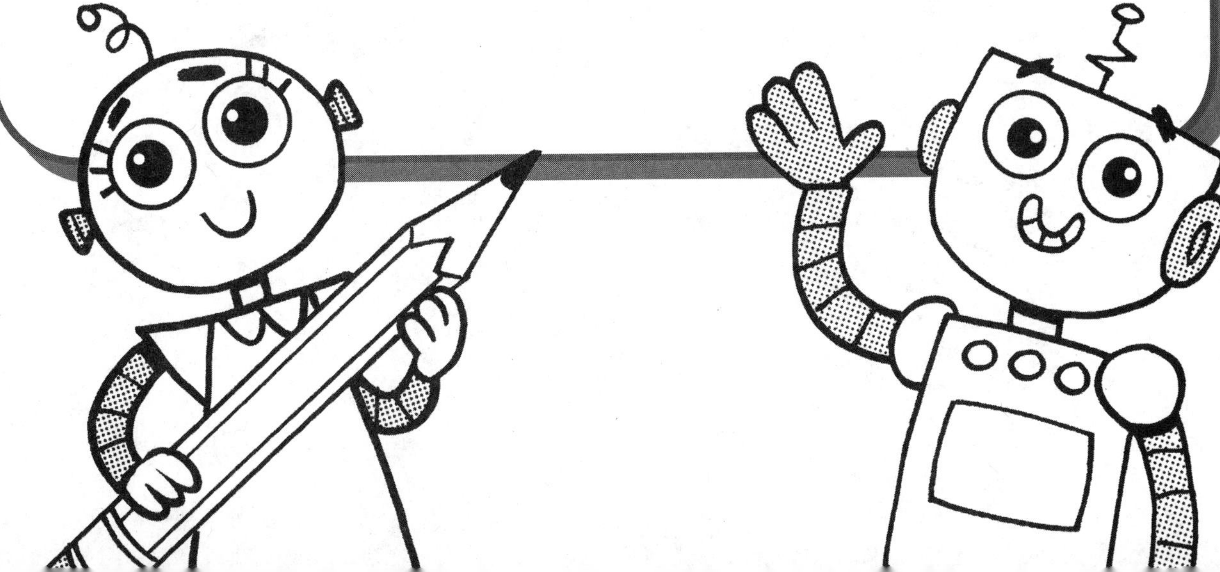

Introduction

Welcome to *Little Learner Packets: Numbers*! The 10 learning packets in this book provide fun, playful activities that teach and reinforce numbers from *1* to *100*. The design, organization, and predictable format of the packets let children complete the pages independently and at their own pace—in school or at home.

Each packet targets a set of numbers for children to practice counting, matching, and tracing. In addition, activities include recognizing and writing numbers and number words, comparing numbers, filling in number sequences, using number lines, and adding one or ten more. A review page at the end of each packet further reinforces the number skills and can be used as a quick and easy way to assess children's learning.

The final packet—Numbers Fun Pack—lets children continue to give their number skills a workout by doing fun activities, such as coloring pictures following a color code, connecting dots to reveal a mystery picture, completing number sequences, and filling in number paths. These pages can also be used as fun assessment tools to gauge children's learning. While the packets are designed to boost number skills, the activities also provide lots of opportunities for children to refine their fine-motor and visual-discrimination skills.

You can use the packets in a variety of ways and with children of all learning styles. Children can complete the activities at their seats or in a learning center. Or they can use the pages as take-home practice. The packets are ideal for encouraging children to work independently and at their own pace. A grid on the introduction page of each packet lets children track their progress as they complete each page. Best of all, the activities support children in meeting the Mathematics standards for Kindergarten. (See the Connections to the Standards box.)

Connections to the Standards

Counting & Cardinality

- Know number names and the count sequence.
- Count to tell the number of objects.
- Compare numbers.

How to Use the Numbers Packets

Copy a class supply of the eight pages for the numbers packet you want to use. Then sequence and staple each set of pages together and distribute the packets to children. All they need to complete the pages are pencils and crayons. TIP: To save paper, we suggest you make double-sided copies.

The format of the learning packets makes the pages easy to use. Packets 1–7 follow the same format. In packets 8 and 9, you'll find many of the same types of activities, but not in the same order as in the first seven packets. Here's what you'll find on each page of packets 1–7:

Page 1: This page introduces the numbers featured in the packet. Children trace the numbers on the page. When the activity is finished, children color in the first box in the tracking grid at the bottom. As they complete each of the remaining pages in the packet, children will color in the corresponding box in this grid.

Pages 2, 3, and 4: Children practice counting, matching, and making sets for the target numbers on these pages. They also practice their number recognition skills.

Page 5: The top section of this page gives children practice in sequencing numbers. At the bottom, they compare sets to identify which has more.

Page 6: On this page, children compare sets to identify which has less. Then they use a number line to find one more and complete simple addition problems by adding 1.

Page 7: This page provides practice in tracing number words and matching them to the corresponding number.

Page 8: These activities review the number skills that have been introduced throughout the packet. Children make associations between each number and number word, create sets, and compare quantities.

TIP: You might use this page as a mini-assessment to check children's progress.

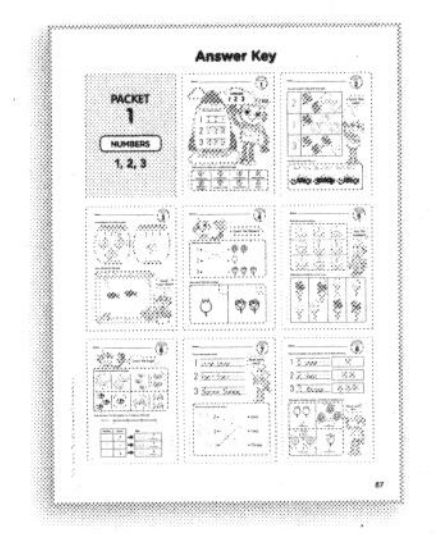

Answer Key: The answer key on pages 87–96 allows you to check children's completed pages at a glance. You can then use the results to determine areas in which they might need additional instruction or practice.

Teaching Tips

Use these tips to help children get the most from the learning packets.

- ★ **Provide a model:** Demonstrate, step by step, how to complete each page in the first packet. Children should then be able to complete the remaining packets independently.
- ★ **Focus on the target number:** Display a set of objects for each target number. Have children count the objects and find the corresponding number card.
- ★ **Promote visual skills:** Have children look carefully at the shape and form of each number they work with to note the similarities and differences in how the numbers are formed.
- ★ **Give number-sequencing practice:** Invite children to sequence a series of three number cards. You might also write number sequences on the board, leaving out one number for children to fill in. For auditory practice, say a few numbers in sequence, such as *1, 2, 3*. Then pause to give children time to name the number that comes next (*4*).

Learning Centers

You might label a separate folder with each child's name and place the packets in the folder to keep in a learning center. Then children can retrieve the assigned packet and work independently through the pages during center time. To make the packets self-checking, you can enlarge the answer keys for each packet, cut apart the images, then sequence and staple them together to create a mini answer key for that packet. Finally, place all of the answer keys in the center. Children can use the answer keys to check their pages as they complete each packet.

Ways to Use the Numbers Learning Packets

Children can work through the packets at their own pace, tracking their progress as they complete each page. The packets are ideal for the following:

- ★ Learning center activity
- ★ Independent seatwork
- ★ One-on-one lesson
- ★ Morning starter
- ★ End of the day wrap-up
- ★ Take-home practice

Assessing Learning

The last pages of the Numbers Packets 1–9 are review pages that can be used to check children's number learning. The Numbers Fun Pack (packet 10) also provides a creative way to assess children's skills. In addition, you might do the following to check children's progress:

- ★ Call out a number and have children create a set of objects for that number.
- ★ Ask children to match the corresponding number and number word cards to a set of objects.
- ★ Choose three or more number cards in a number sequence. Place the cards in random order, then have children put the cards in the correct order.
- ★ Call out numbers at random and have children write them on paper or on the board.

Name: ____________________

Color each box when you complete that page.

① Introduction	② Count & Color	③ Count & Draw	④ Match & Write
⑤ Sequence & Compare	⑥ Compare & Add	⑦ Trace & Match	⑧ Review

Name: ______________________________

Say each number. Color that many balls.

Use the code to color the cars.

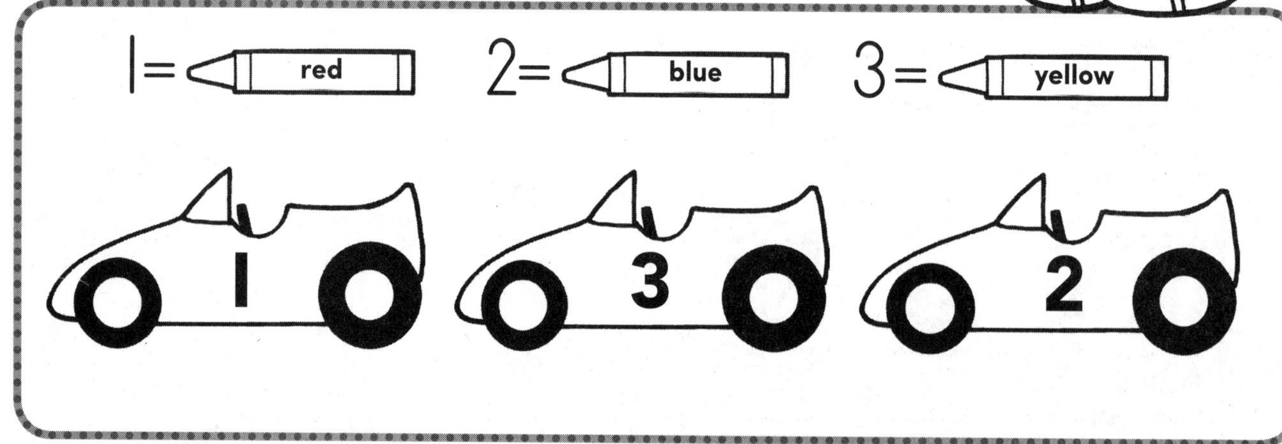

Name: ______________________________

Count the fish. Circle the number.

Draw 2 fish in the fish tank.

Name: ______________________________

Count the flowers!

Match each number to its set.

1 • •  

2 • • 

3 • •   

How many? Write the number.

Name: ______________________________

Write the missing numbers.

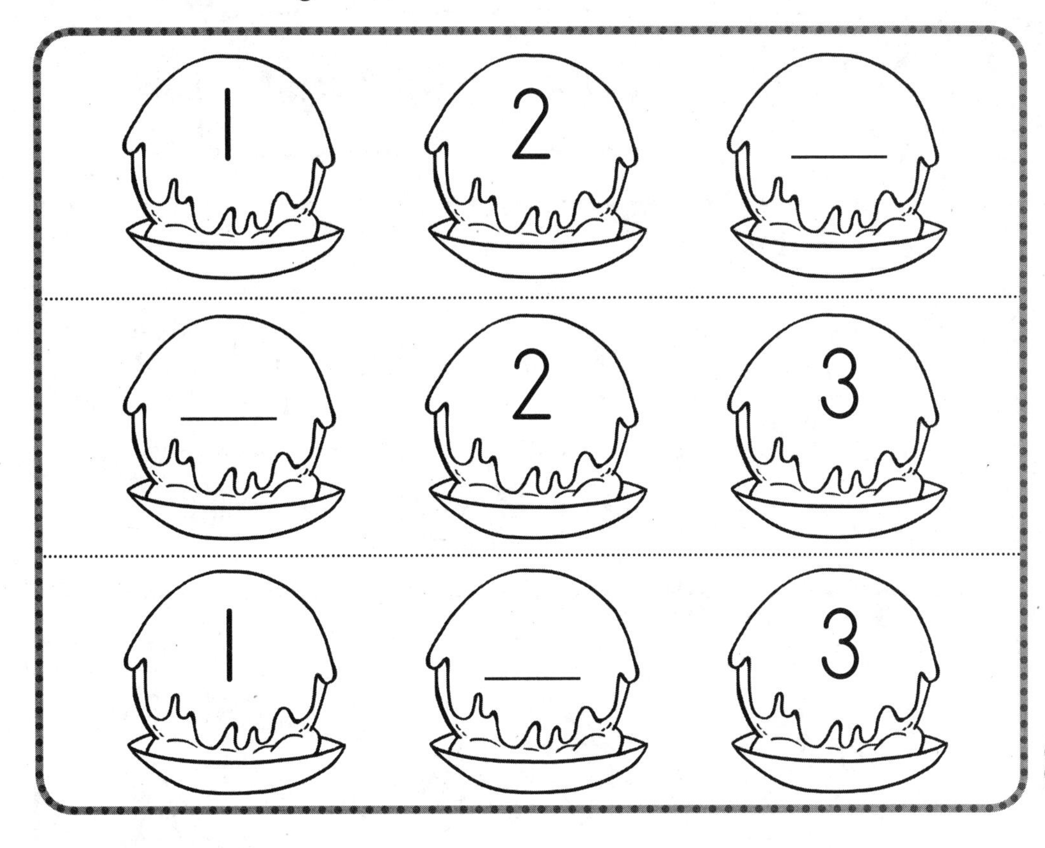

Color each cone that has more scoops.

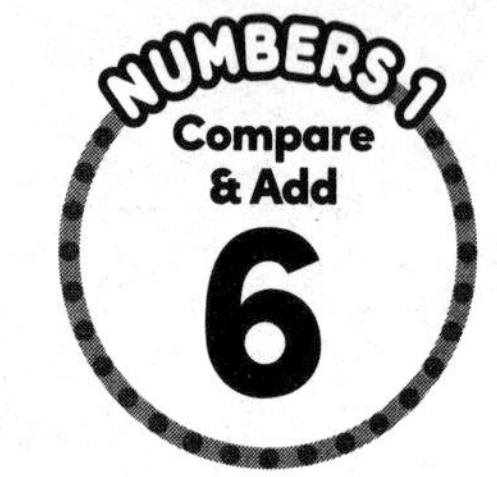

Name: ______________________________

Count the bugs!

Circle each set that has fewer bugs.

Write one more. Use the number line to help you. Then add.

START HERE. 0 — 1 — 2 — 3

Number	1 more
1	
0	
2	

Add.

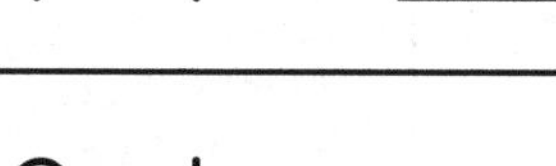

1 + 1 = ______

0 + 1 = ______

2 + 1 = ______

Name: ____

Trace each number word.

1 

2 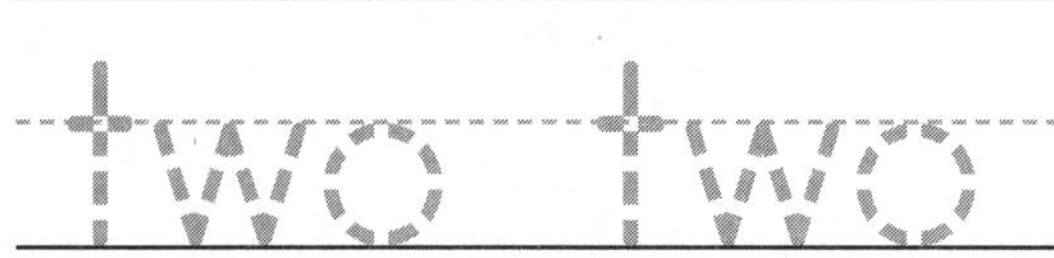

3 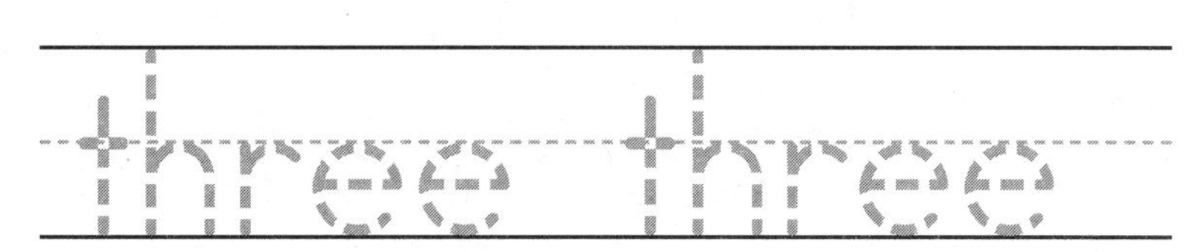

Read each word!

Match each number to its name.

 2 • • one

 3 • • two

 1 • • three

Name: ______________________________

Trace each number and word. Draw a set of balls in the box.

1 1 one

2 2 two

3 3 three

How many? Write the number. Circle the set that has more.

Name: ______________________________

Color each box when you complete that page.

① Introduction	② Count & Color	③ Count & Draw	④ Match & Write
⑤ Sequence & Compare	⑥ Compare & Add	⑦ Trace & Match	⑧ Review

Name: ________________________________

Say each number. Color that many toys.

5	
6	
4	

Use the code to color the cars.

4 = red 5 = blue 6 = yellow

5 6 4

Name: ______________________________

Count the fish. Circle the number.

Draw 4 fish in the fish tank.

Name: ______________________________

Count the fruits!

Match each number to its set.

4 • • 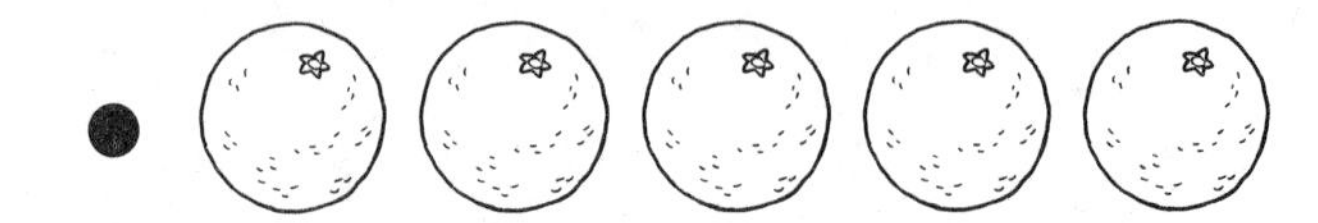

6 • • 

5 • • 

How many? Write the number.

Name: ____

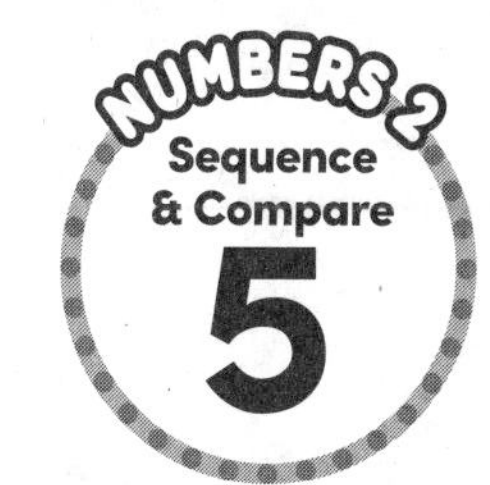

Write the missing numbers.

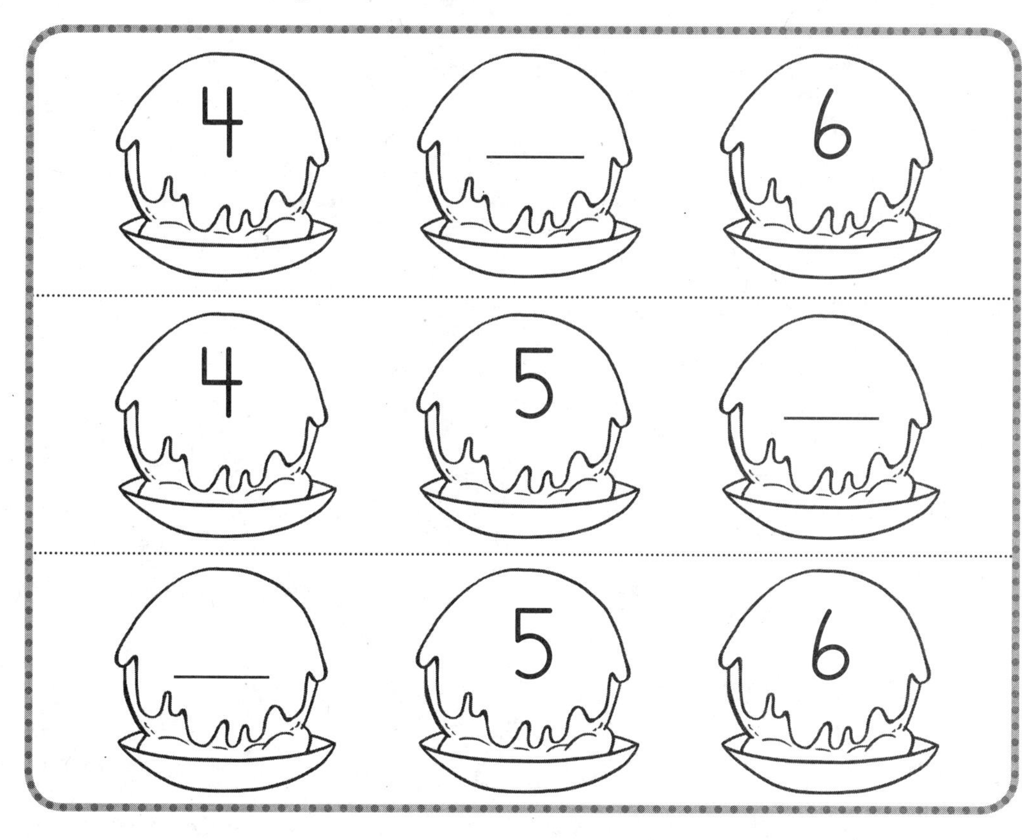

Color each sundae that has more chocolate chips.

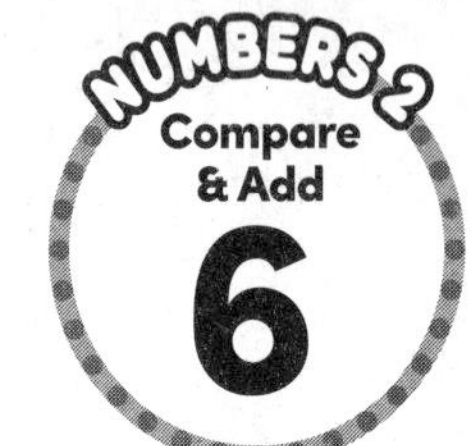

Name: ____

Count the bugs!

Circle each set that has fewer bugs.

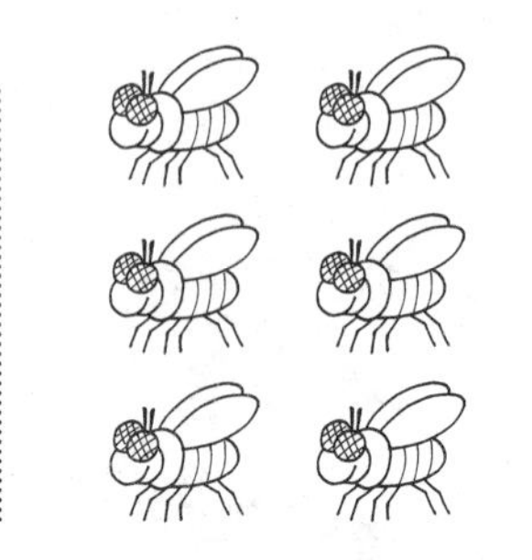

Write one more. Use the number line to help you. Then add.

START HERE.

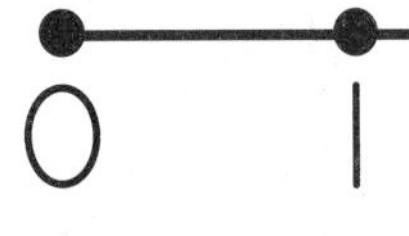

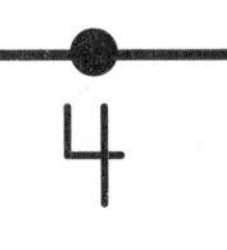

0 1 2 3 4 5 6

Number	1 more
4	
3	
5	

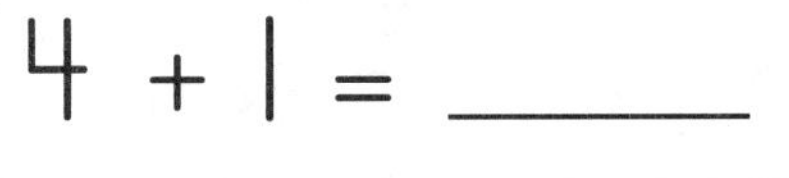
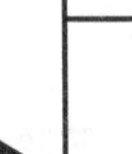
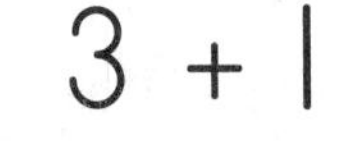

Add.

4 + 1 = ______

3 + 1 = ______

5 + 1 = ______

Name: ______________________________

Trace each number word.

4 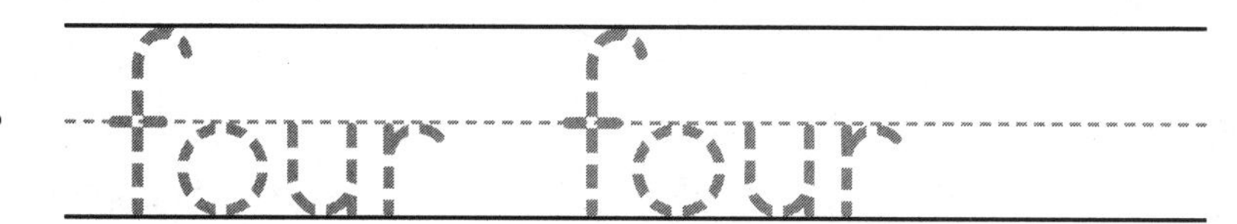

5 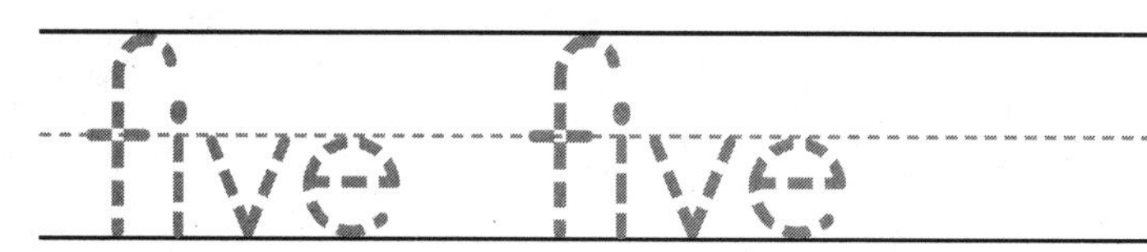

6 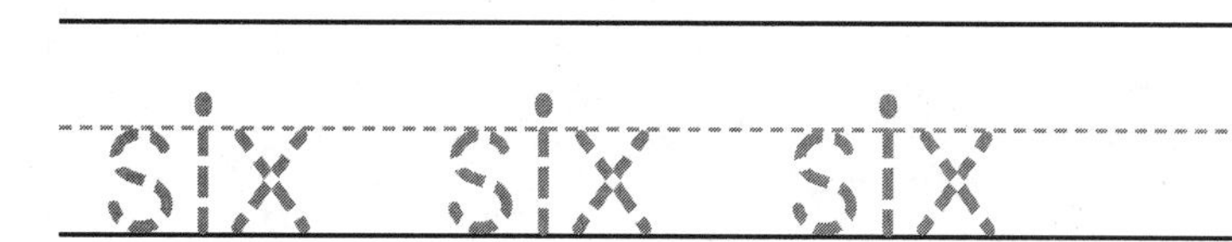

Match each number to its name.

 5 • • five

 6 •  • four

 4 • • six

Name: ______________________________

Trace each number and word. Draw a set of balls in the box.

4 4 four

5 5 five

6 6 six

How many? Write the number. Circle the set that has more.

Name: ______________________________

NUMBERS 3
Introduction
1

NUMBERS

7 8 9

Hi!

Trace each number.

7 7 7 7

8 8 8 8

9 9 9 9

Color each box when you complete that page.

1 Introduction	2 Count & Color	3 Count & Draw	4 Match & Write
5 Sequence & Compare	6 Compare & Add	7 Trace & Match	8 Review

Name: ______________________________

Say each number. Color that many balls.

7	
8	
9	

Count the balls!

Use the code to color the cars.

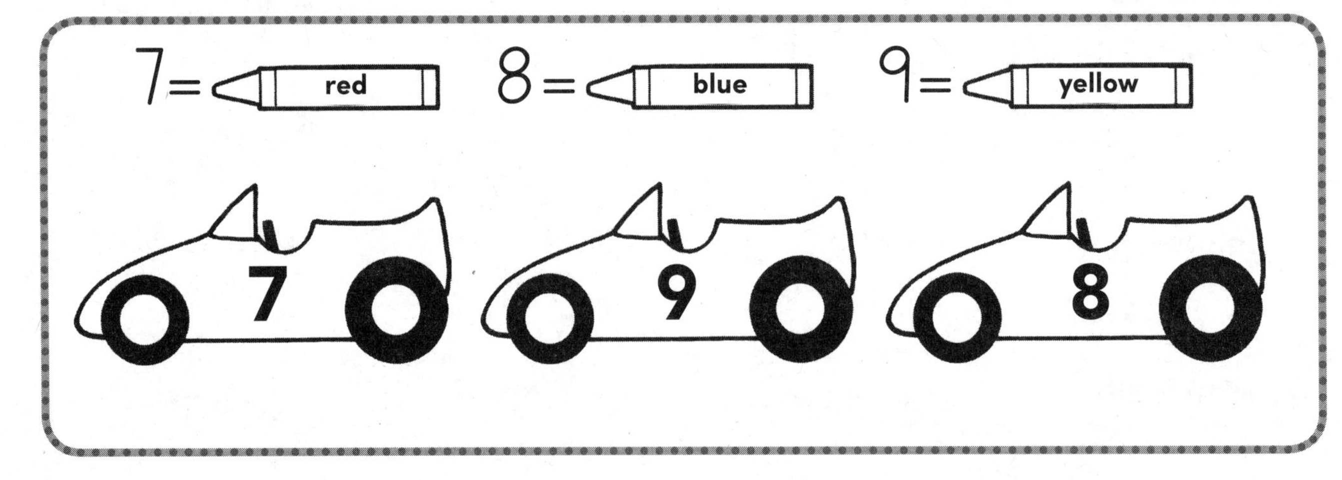

Name: ______________________________

Count the fish. Circle the number.

Draw 8 fish in the fish tank.

Name: ______________________________

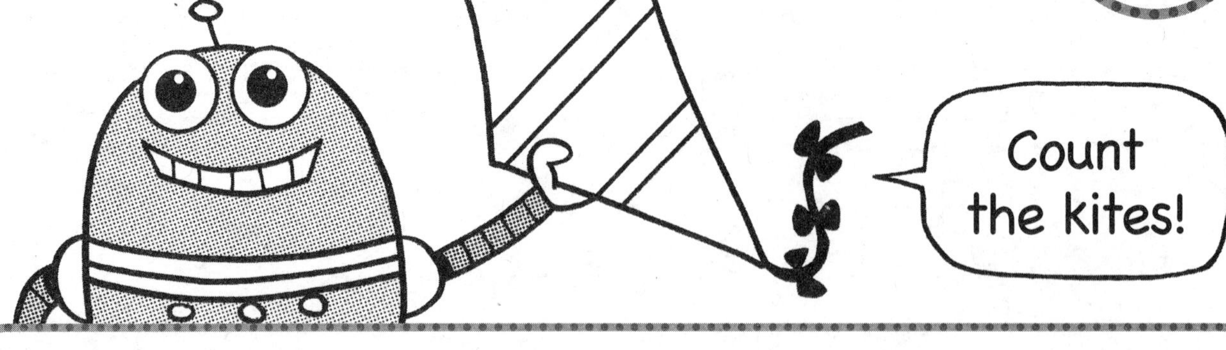

Match each number to its set.

8 •

7 •

9 •

How many? Write the number.

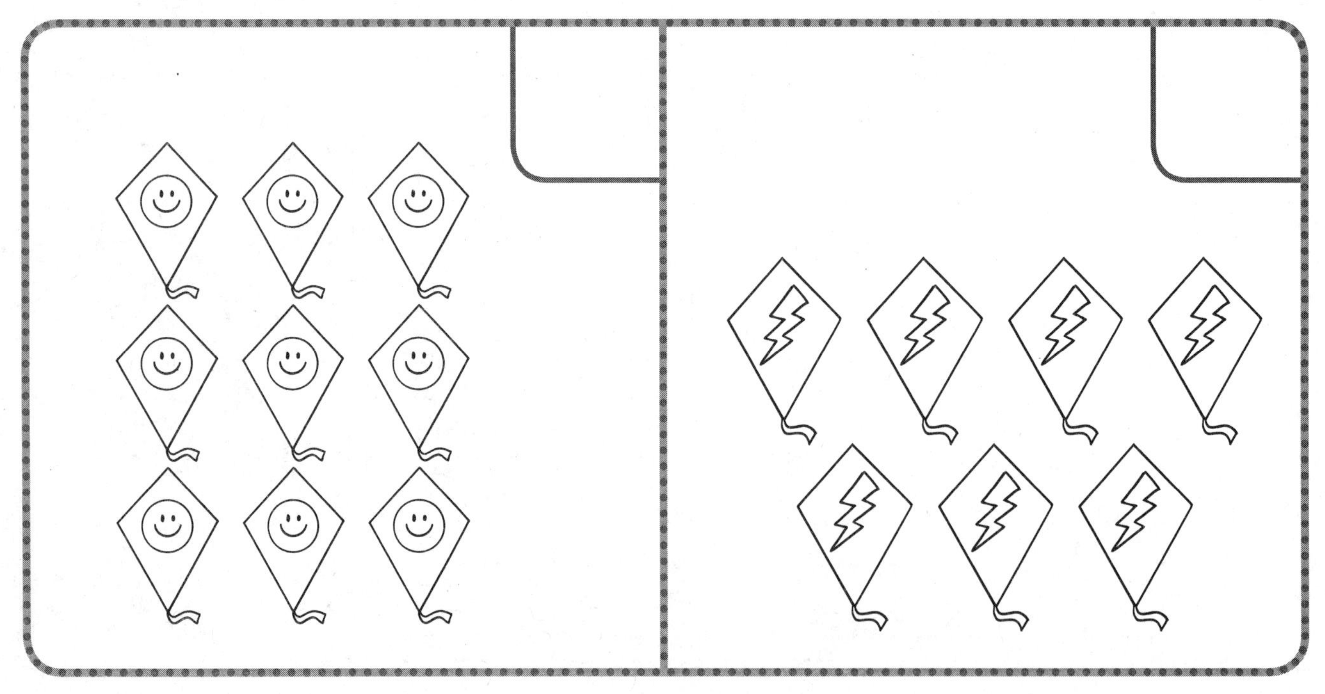

Name: ______________________________

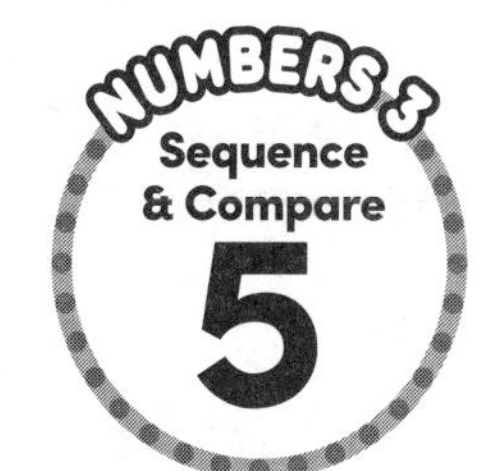

Write the missing numbers.

Color each sundae that has more chocolate chips.

Name: ______________________________

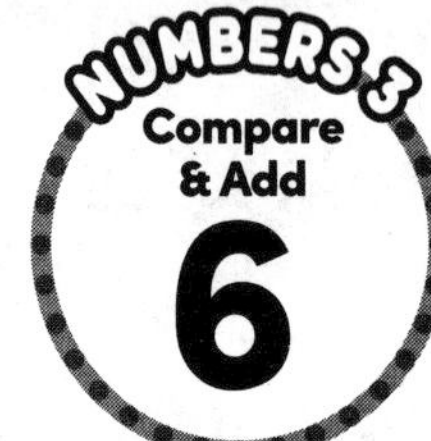

Count the school tools!

Circle each set that has fewer tools.

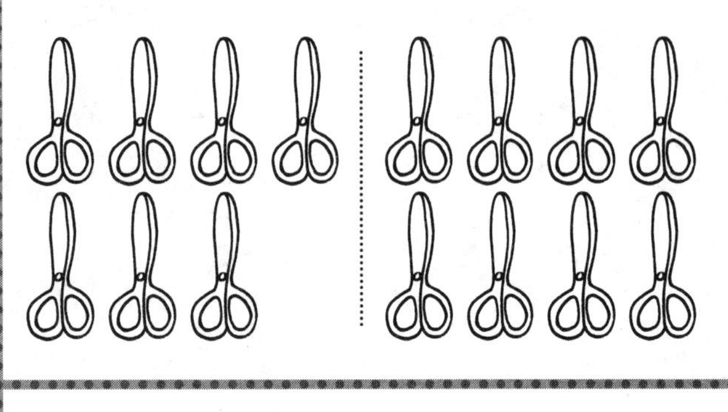

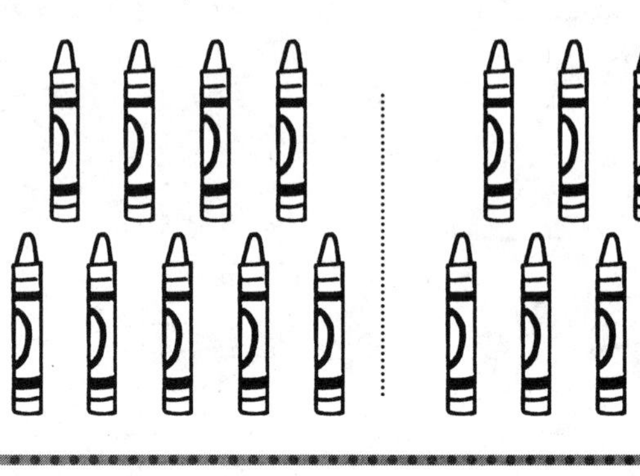

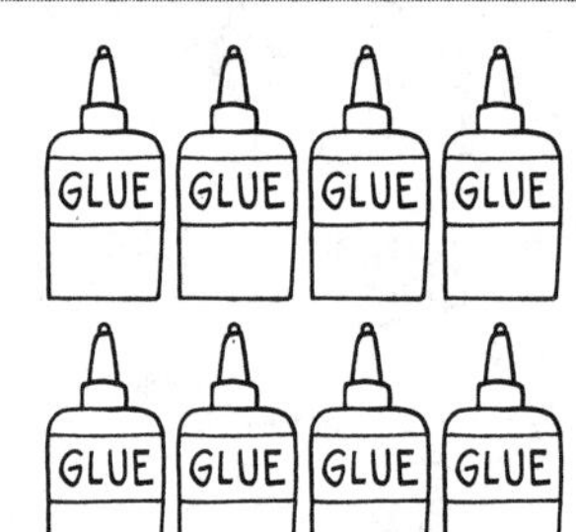

Write one more. Use the number line to help you. Then add.

START HERE.

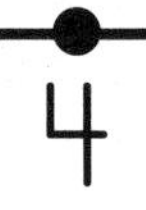

0 1 2 3 4 5 6 7 8 9

Number	1 more		Add.
8		→	8 + 1 = ______
7		→	7 + 1 = ______
6		→	6 + 1 = ______

Name: ______________________________

Trace each number word.

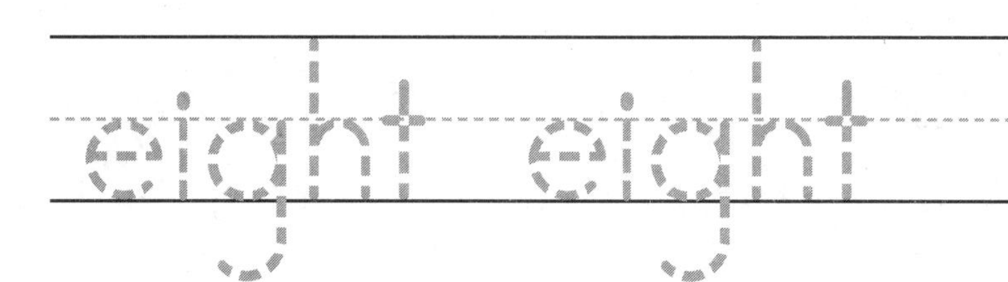

8 eight eight

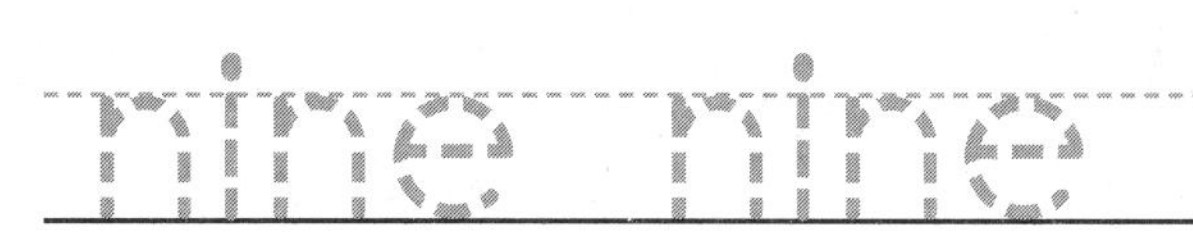

9

Match each number to its name.

8 •	• nine
9 •	• seven
7 •	• eight

Name: ______________________________

Trace each number and word. Draw a set of balls in the box.

7 7 seven

8 8 eight

9 9 nine

How many? Write the number. Circle the set that has more.

Name: ______________________________

NUMBERS 4
Introduction
1

NUMBERS
10 11 12

Hi!

Trace each number.

10

11

12

Color each box when you complete that page.

① Introduction	② Count & Color	③ Count & Draw	④ Match & Write
⑤ Sequence & Compare	⑥ Compare & Add	⑦ Trace & Match	⑧ Review

Name: ______________________________

Say each number. Color that many toys.

11	
10	
12	

Count the toys!

Use the code to color the cars.

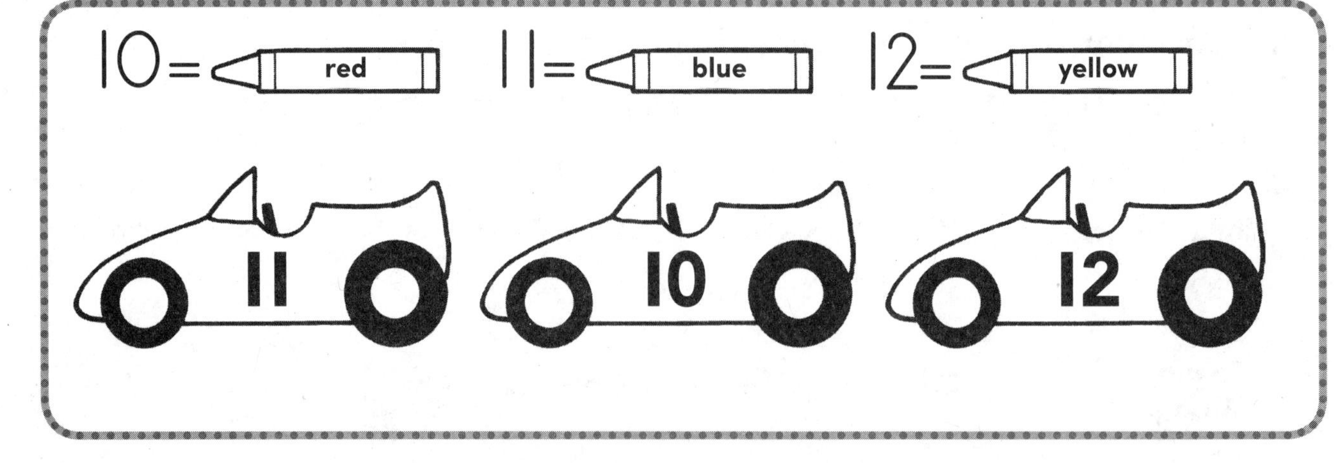

Name: ___________________________

Count the fish. Circle the number.

Draw 11 fish in the fish tank.

Name: ___________________________________

Count the pets!

Match each number to its set.

12 •

10 •

11 •

How many? Write the number.

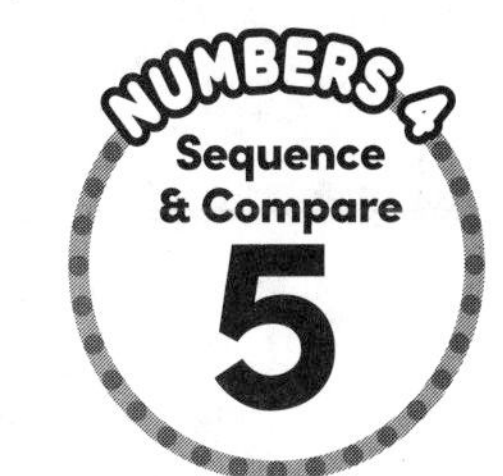

Name: ____

Write the missing numbers.

Color each sundae that has more chocolate chips.

Name: ______________________________

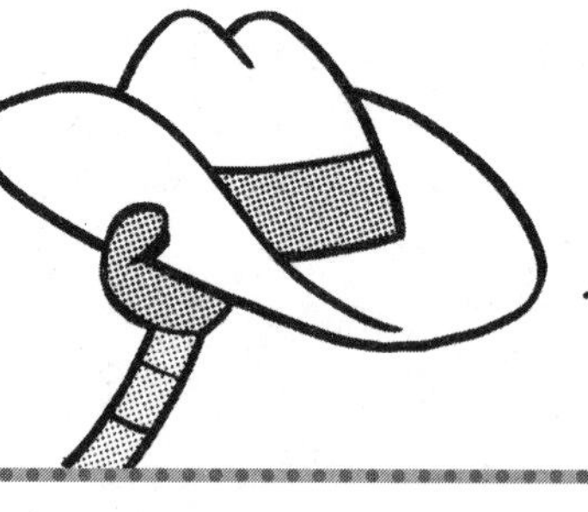

Count the hats!

Circle the set that has fewer hats.

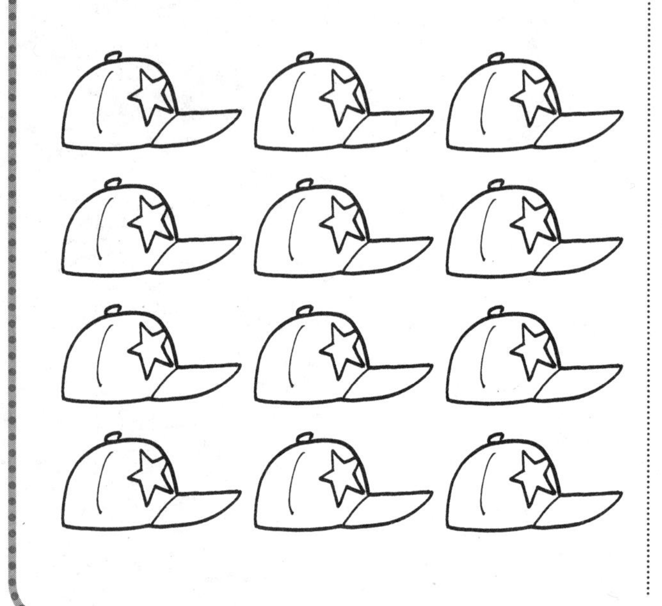

Write one more. Use the number line to help you. Then add.

START HERE.

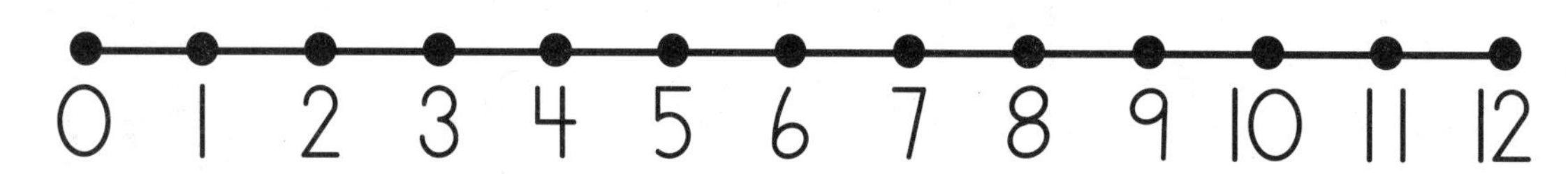

Number	1 more
10	
9	
11	

Add.

10 + 1 = ______

9 + 1 = ______

11 + 1 = ______

Name: ____________________

Trace each number word.

10 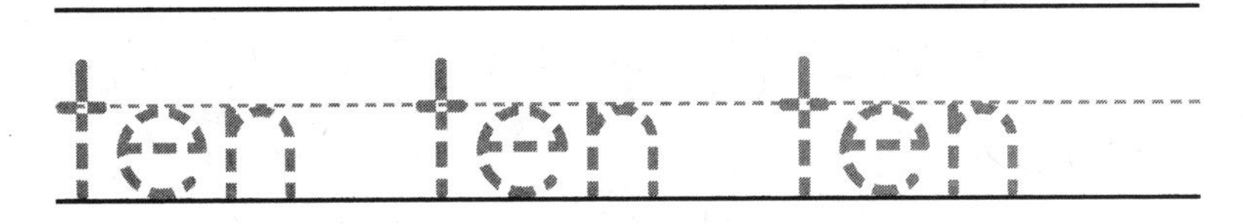

11 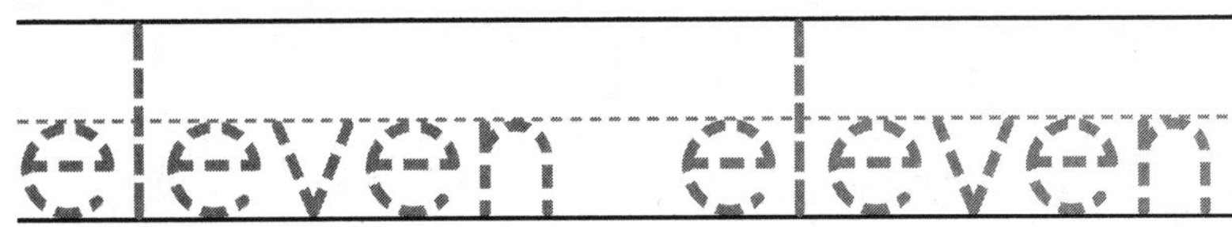

12 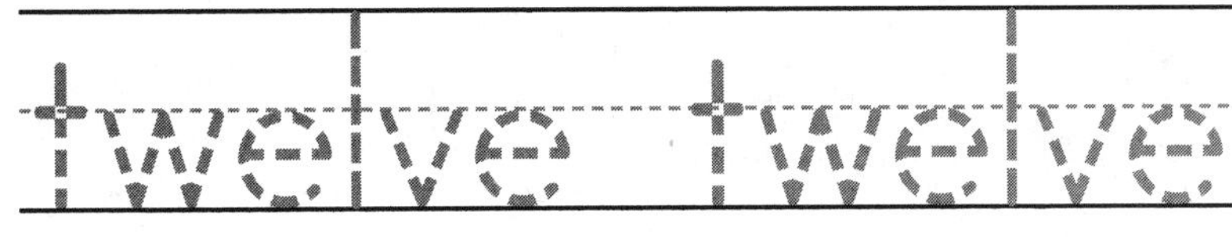

Read each word!

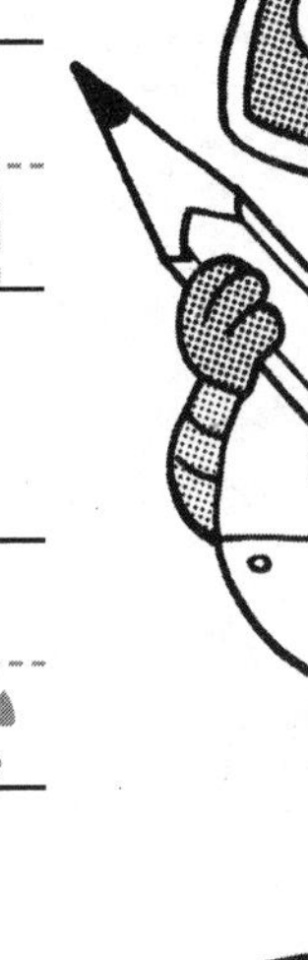

Match each number to its name.

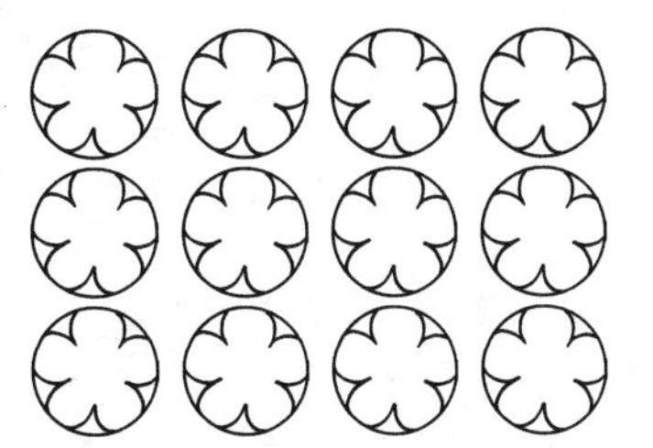

12 •

• ten

10 •

11 •

• eleven

Name: ______________________________

Trace each number and word. Draw a set of balls in the box.

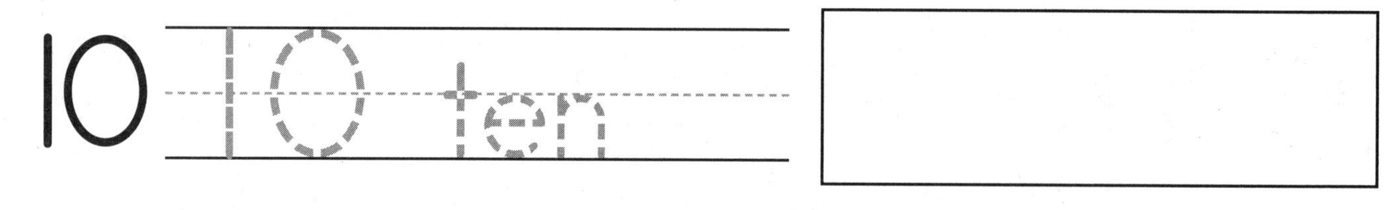

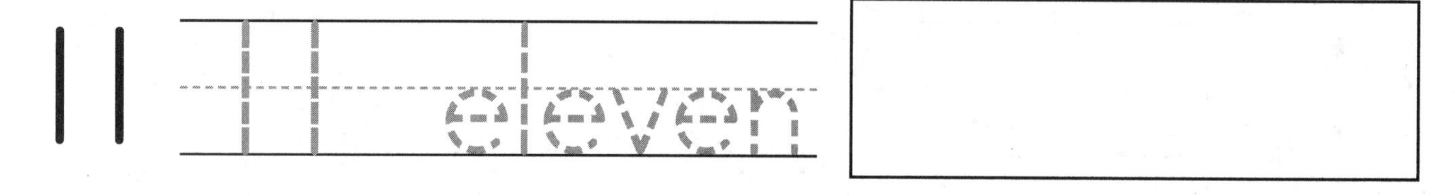

12 12 twelve

How many? Write the number. Circle the set that has more.

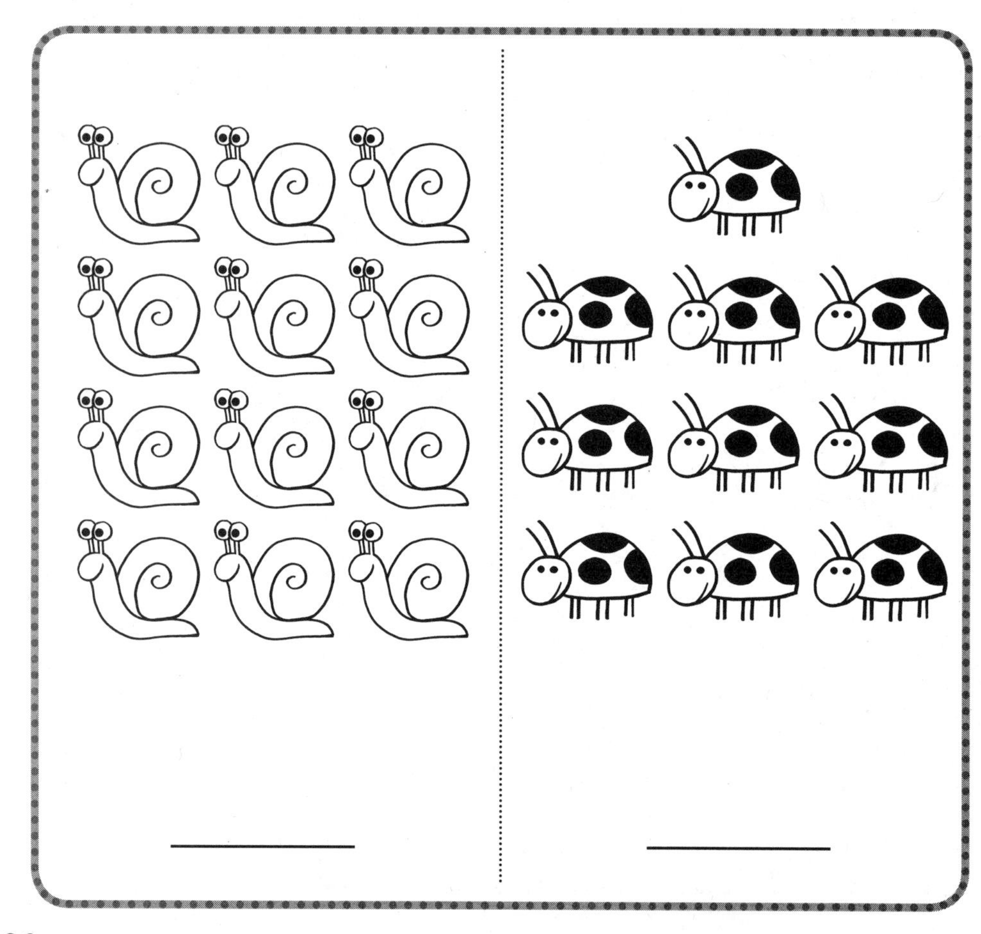

Name: ___________________________

Color each box when you complete that page.

1 Introduction	2 Count & Color	3 Count & Draw	4 Match & Write
5 Sequence & Compare	6 Compare & Add	7 Trace & Match	8 Review

NUMBERS 5
Count & Color
2

Name: ______________________________

Say each number. Color that many balls.

13	
14	
15	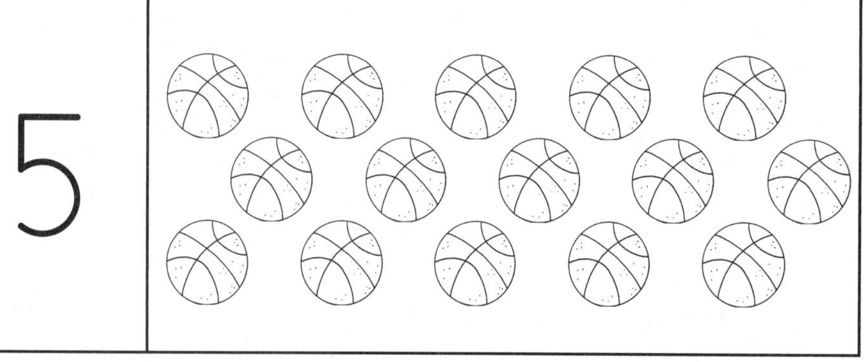

Use the code to color the cars.

Name: ______________________________________

Count the fish. Circle the number.

Draw 14 fish in the fish tank.

Name: ______________________________

Match each number to its set.

15 •

13 •

14 •

How many? Write the number.

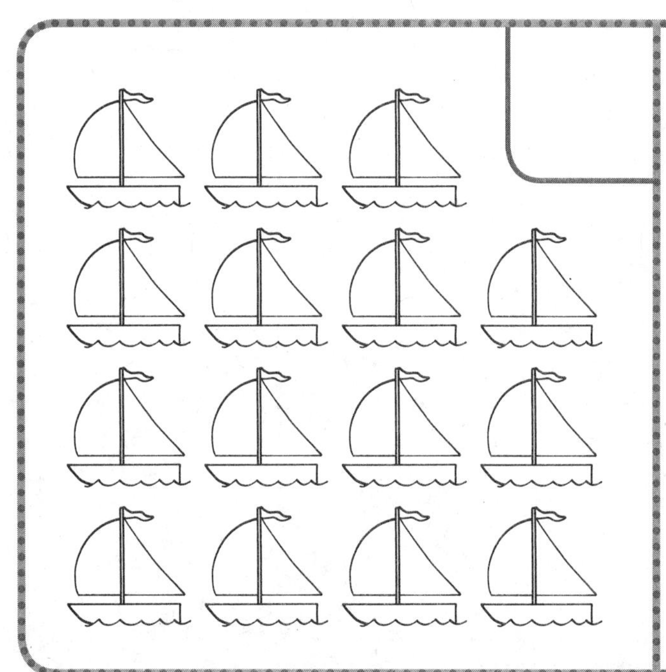

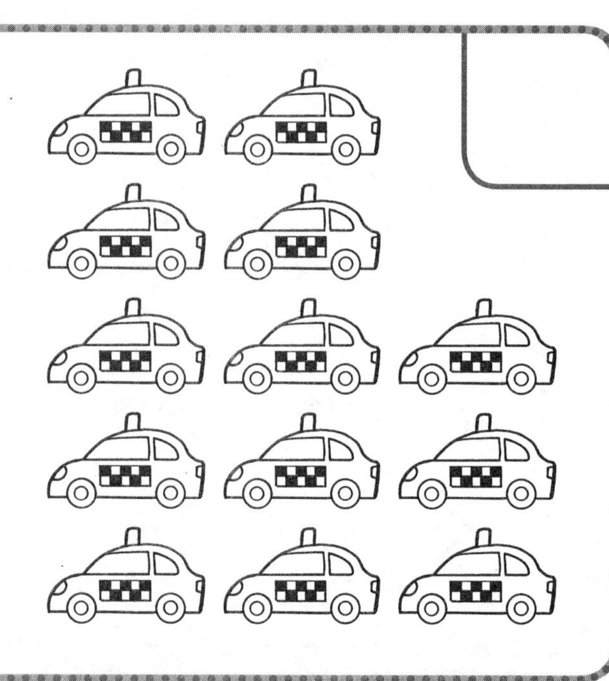

Name: ______________________________

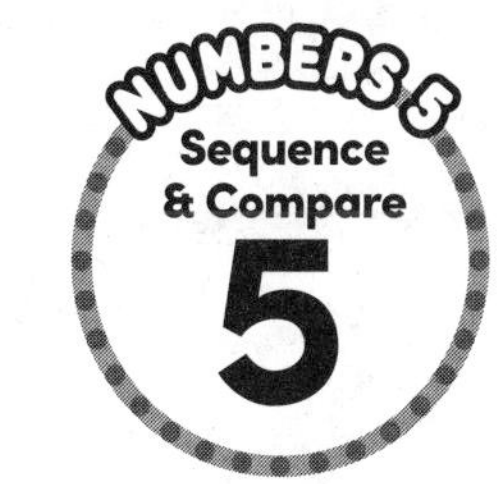

Write the missing numbers.

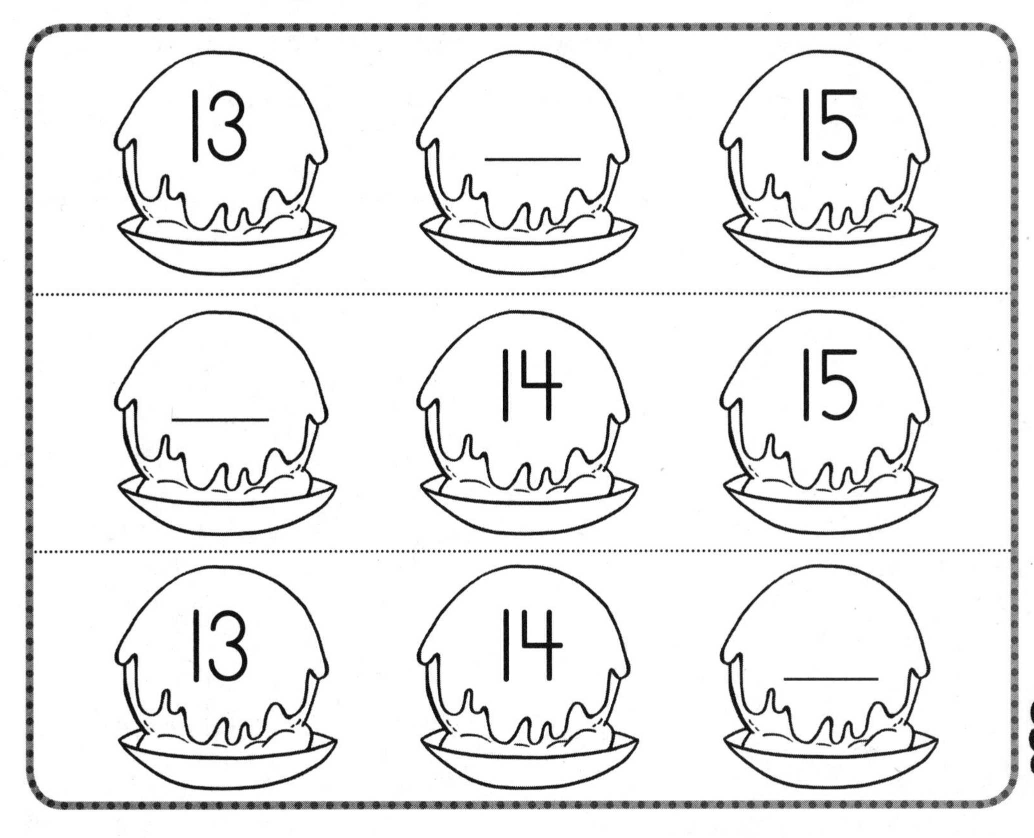

Say the numbers!

Color the sundae that has more chocolate chips.

Name: ____________________________

NUMBERS 5
Compare & Add
6

Count the gifts!

Circle the set that has fewer gifts.

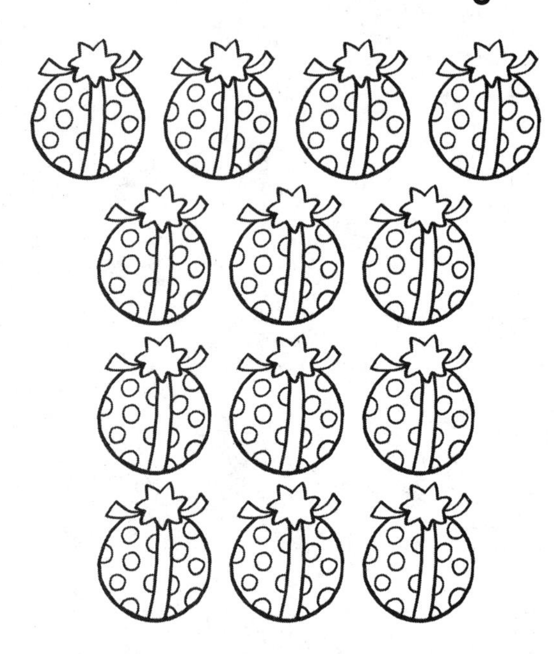

Write one more. Use the number line to help you. Then add.

START HERE.

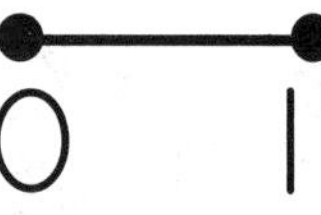

10 11 12 13 14 15

Number	1 more
14	
12	
13	

Add.

$14 + 1 = \_\_\_\_$

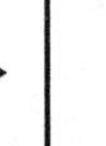

$12 + 1 = \_\_\_\_$

$13 + 1 = \_\_\_\_$

Name: ______________________________

Trace each number word.

13 thirteen

14 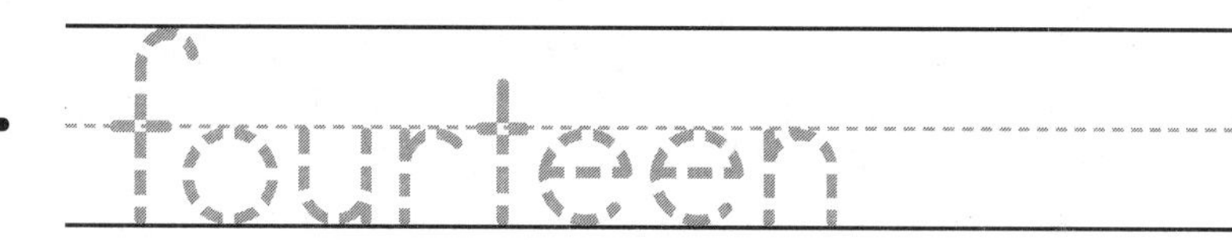

15 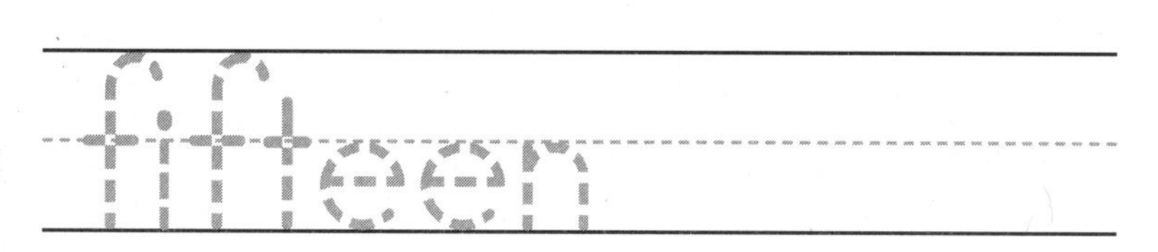

Match each number to its name.

13 • • thirteen

14 • • fifteen

15 • • fourteen

Name: ______________________________

Trace each number and word. Draw a set of balls in the box.

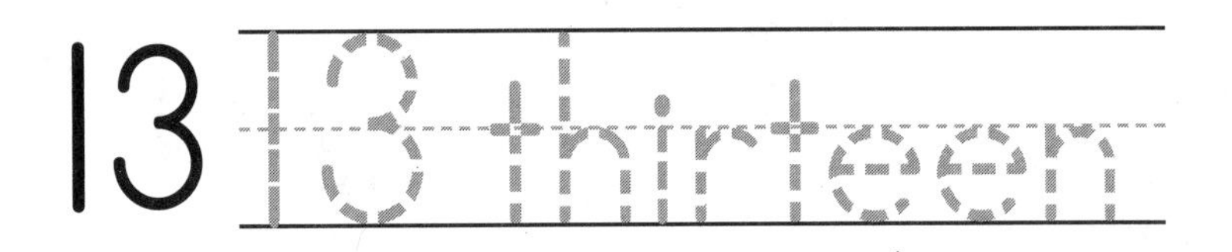

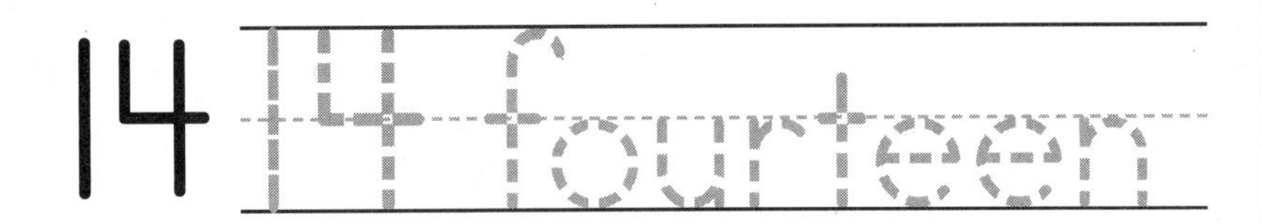

15 15 fifteen

How many? Write the number. Circle the set that has more.

Name: ______________________________

NUMBERS 6

Introduction

1

Color each box when you complete that page.

① Introduction	② Count & Color	③ Count & Draw	④ Match & Write
⑤ Sequence & Compare	⑥ Compare & Add	⑦ Trace & Match	⑧ Review

Name: ____________________

Say each number. Color that many toys.

18	
16	
17	

Use the code to color the cars.

Name: ______________________________

Count the fish. Circle the number.

Draw 17 fish in the fish tank.

Name: ____

Count the foods!

Match each number to its set.

16 •

17 •

18 •

How many? Write the number.

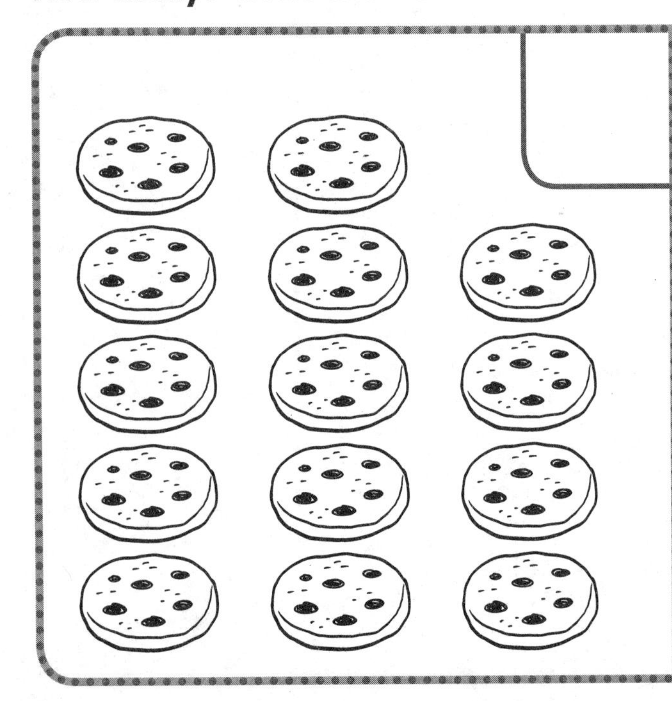

Name: ______________________________

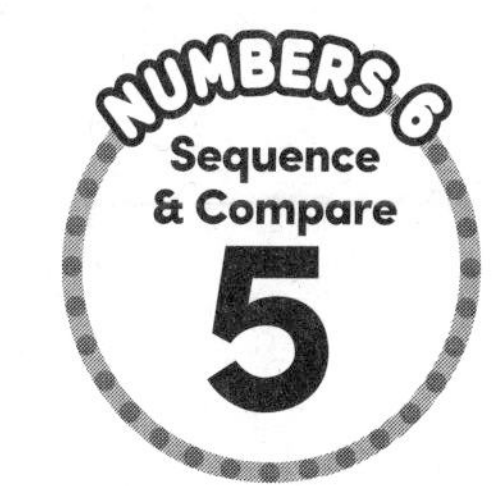

Write the missing numbers.

Color the sundae that has more chocolate chips.

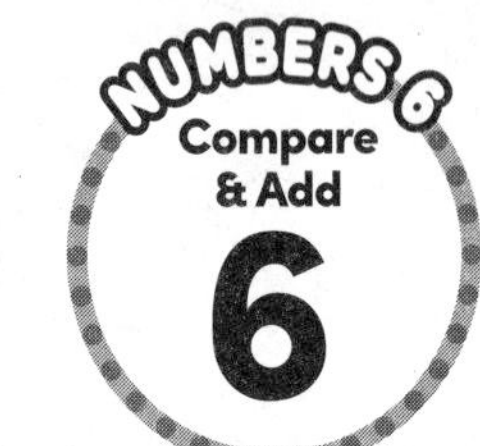

Name: ______________________________

Circle the set that has fewer buttons.

Write one more. Use the number line to help you. Then add.

START HERE. 10 11 12 13 14 15 16 17 18

Number	1 more
16	
15	
17	

Add.

16 + 1 = ______
15 + 1 = ______
17 + 1 = ______

Name: ______________________________

Trace each number word.

16 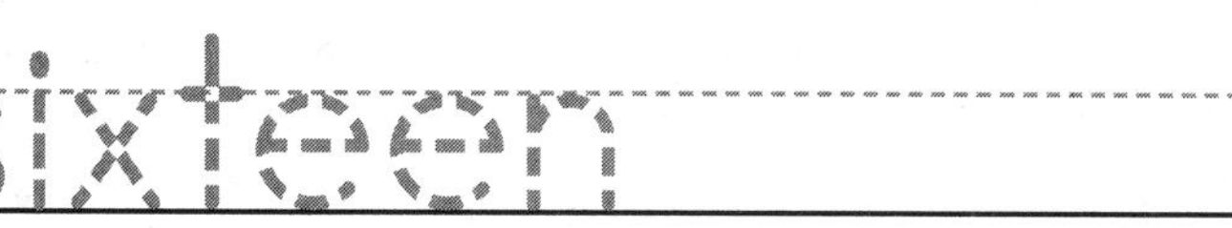

17 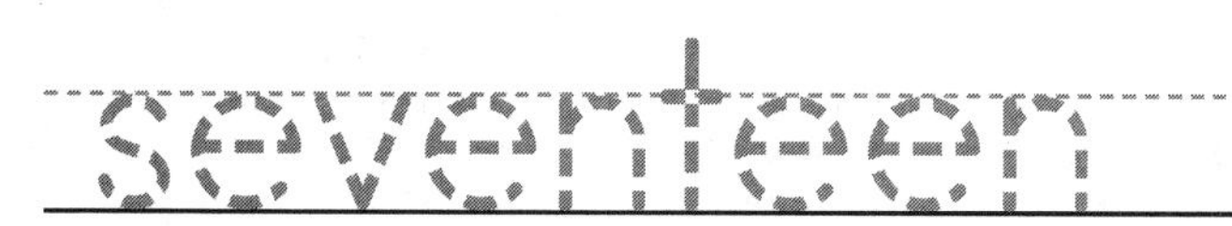

18 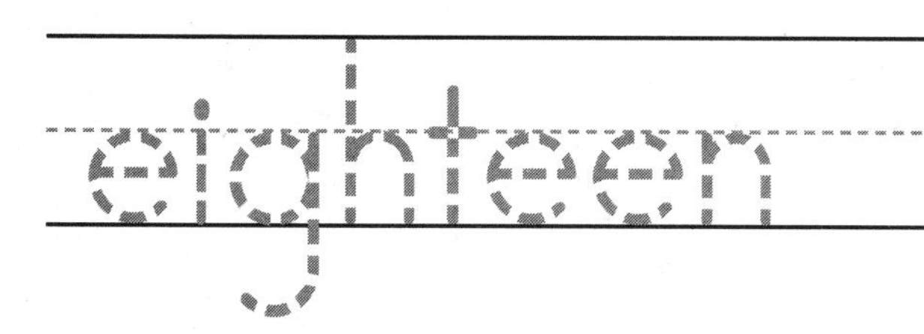

Match each number to its name.

18 •  • sixteen

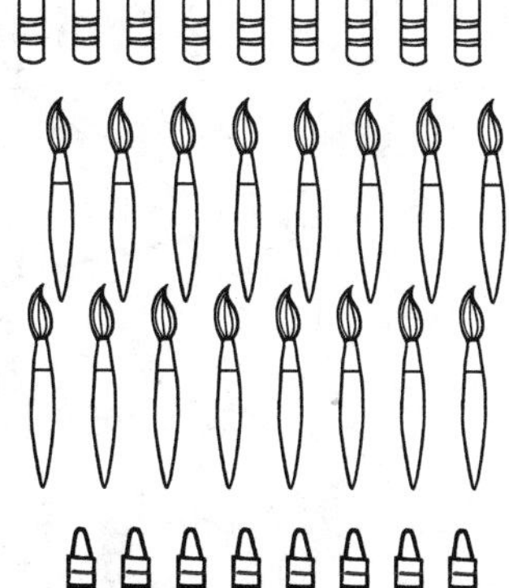

16 • • eighteen

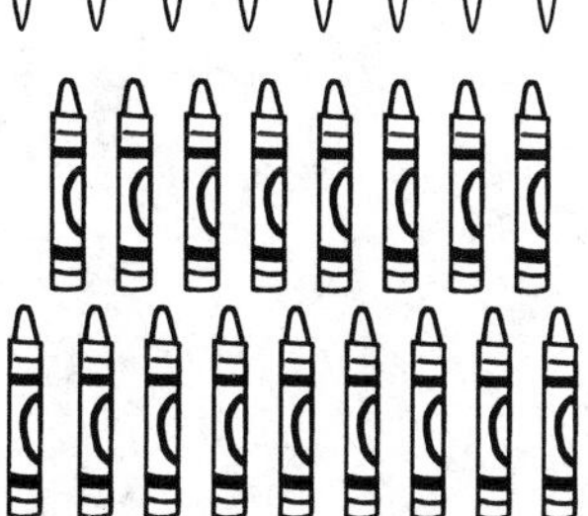

17 • • seventeen

Name: ______________________________

Trace each number and word. Draw a set of balls in the box.

16 16 sixteen

17 17 seventeen

18 18 eighteen

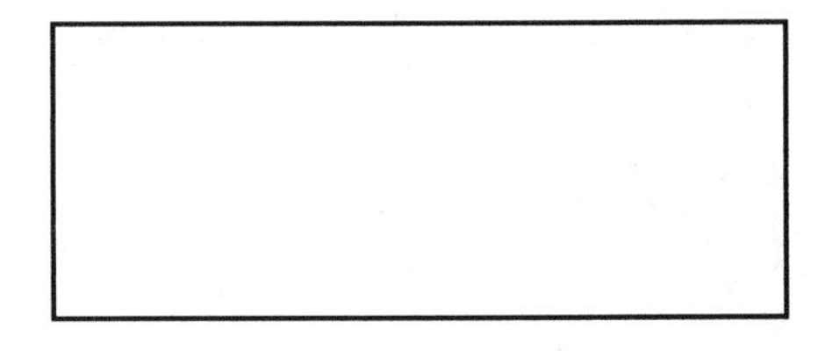

How many? Write the number. Circle the set that has more.

NUMBERS 7
Introduction
1

Name: ____________________

NUMBERS
19 20

Hi!

Trace each number.

19 19 19

20 20 20

Color each box when you complete that page.

① Introduction	② Count & Color	③ Count & Draw	④ Match & Write
⑤ Sequence & Compare	⑥ Compare & Add	⑦ Trace & Match	⑧ Review

Name: ______________________________

Say each number. Color that many animals.

20	
19	

Use the code to color the cars.

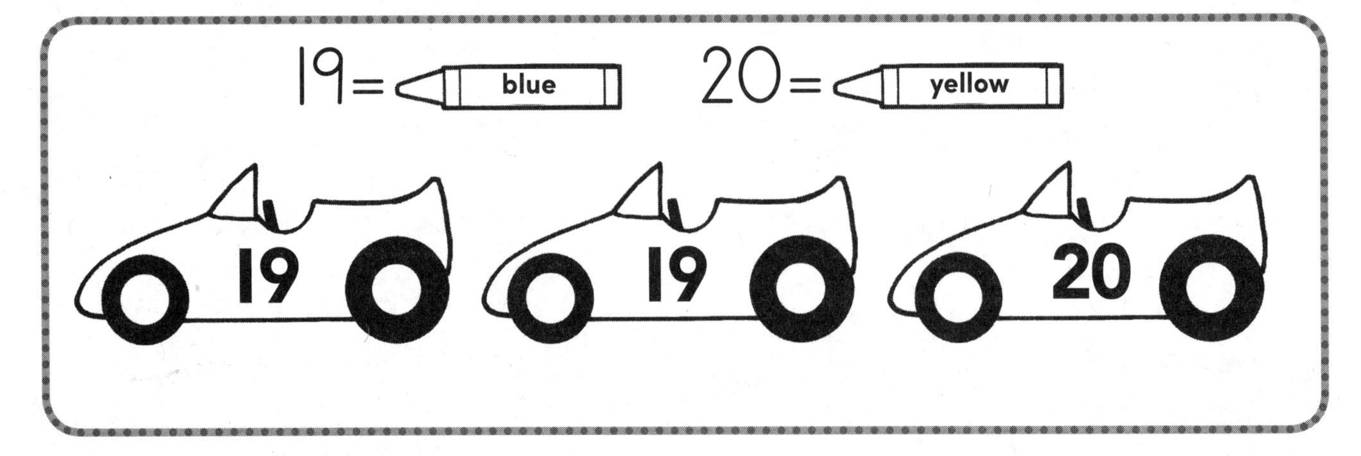

Name: ______________________________

Count the fish. Circle the number.

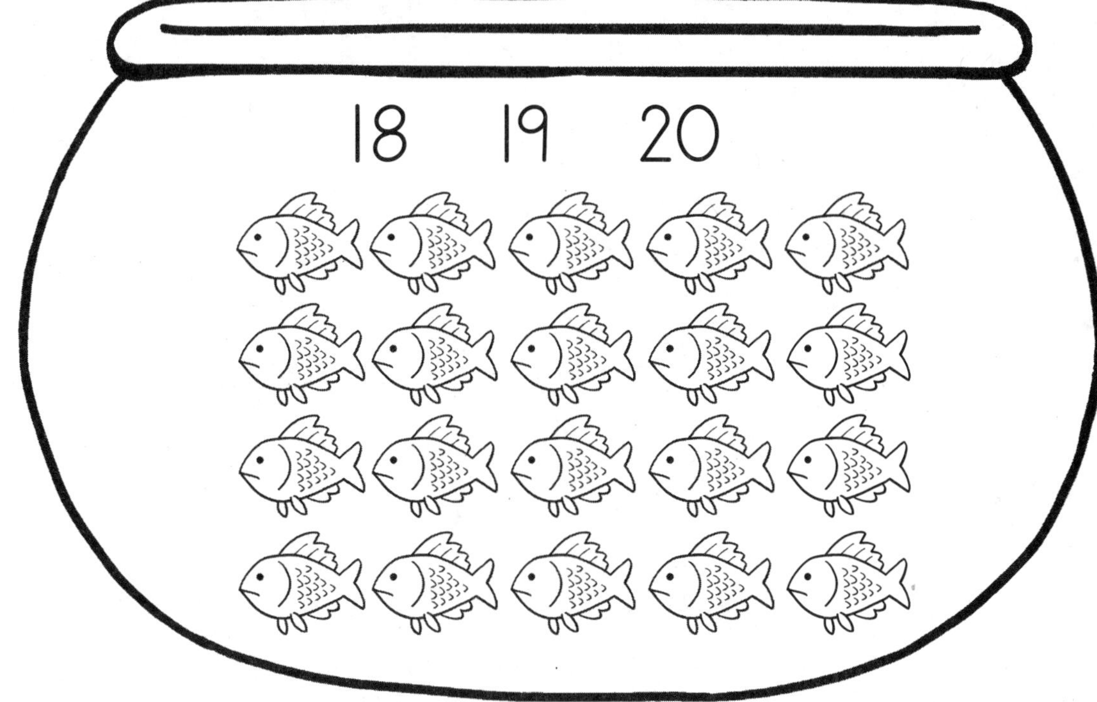

Draw 19 fish in the fish tank.

Name: ______________________________

Match each number to its set.

19 •   •

20 •   •

How many? Write the number.

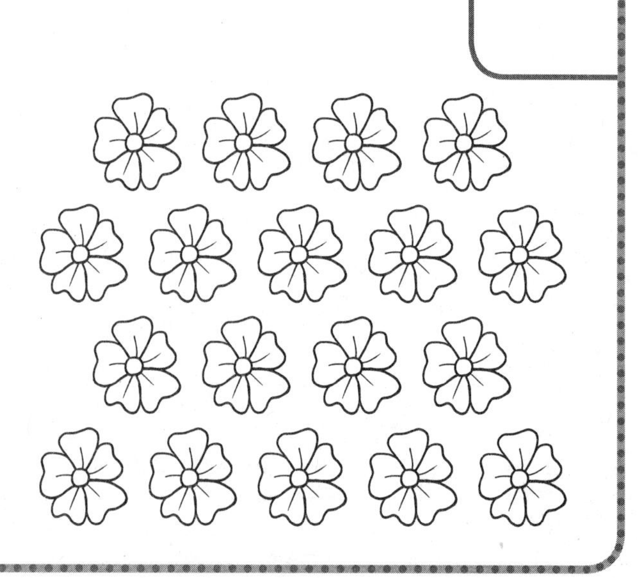

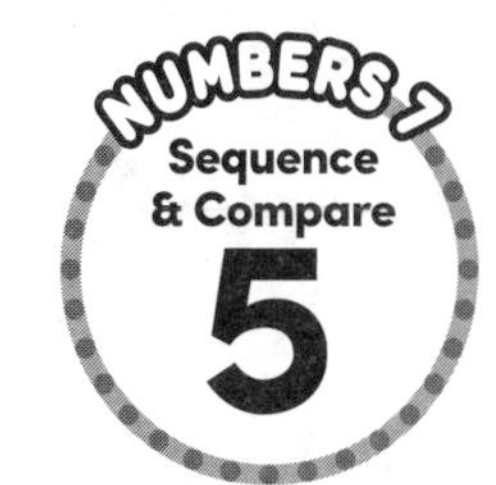

Name: ______________________________

Write the missing numbers.

Color the sundae that has more chocolate chips.

Name: ____________________________________

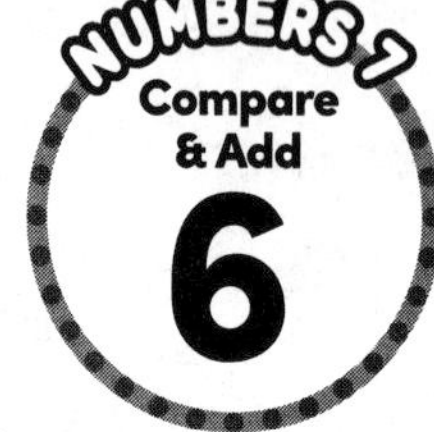

Count the bugs!

Circle the set that has fewer bugs.

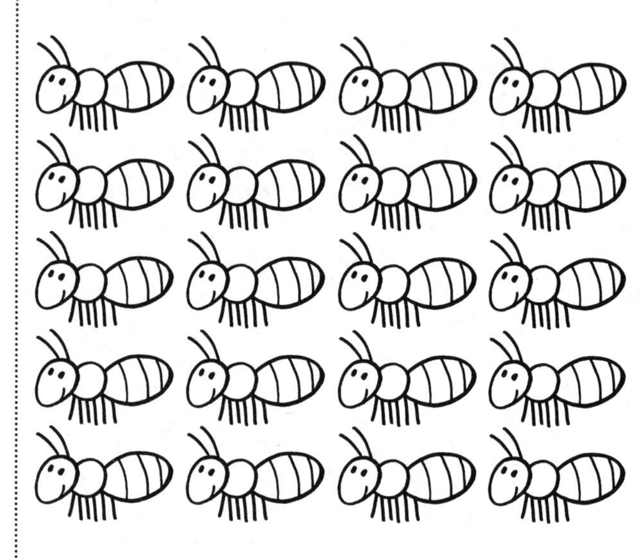

Write one more. Use the number line to help you. Then add.

START HERE.

Number	1 more		Add.
18			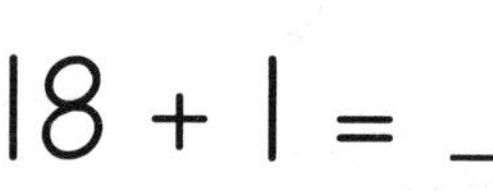 18 + 1 = ______
17		→	17 + 1 = ______
19			19 + 1 = ______

Name: ______________________________

Trace each number word.

19
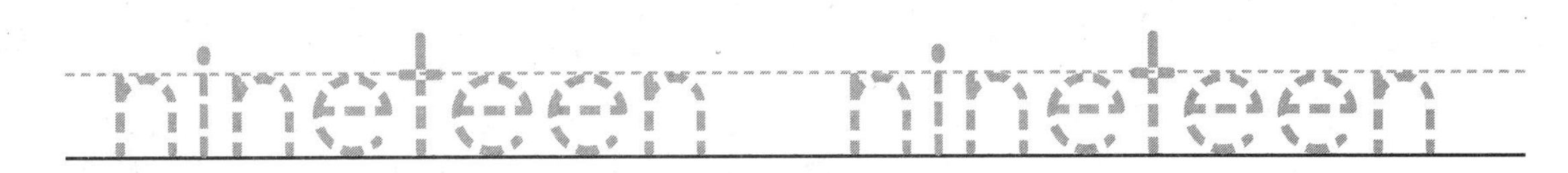

20 twenty twenty

Match each number to its name.

19 • • twenty

20 • • nineteen

Name: ____________________

Trace each number and word. Draw a set of balls in the box.

19 19 nineteen

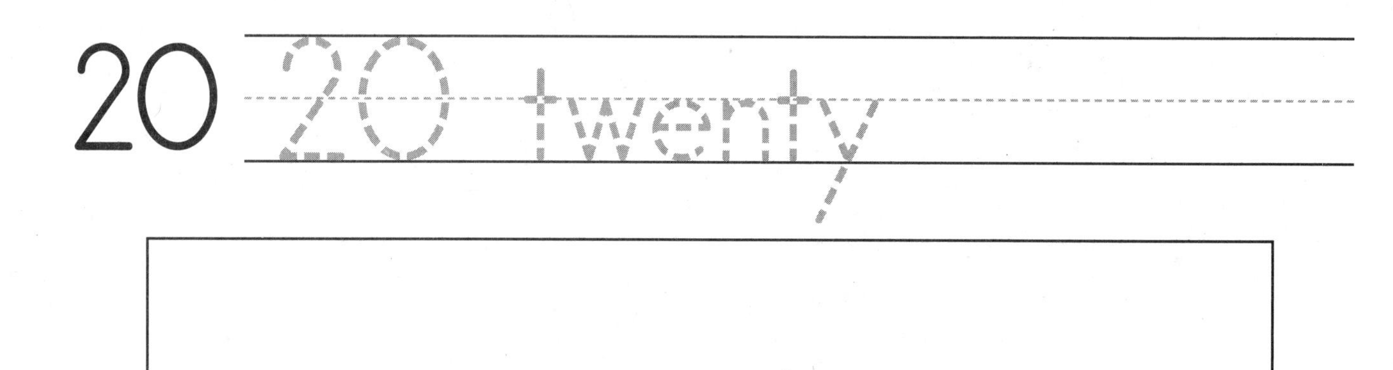

20 20 twenty

How many? Write the number. Circle the set that has fewer.

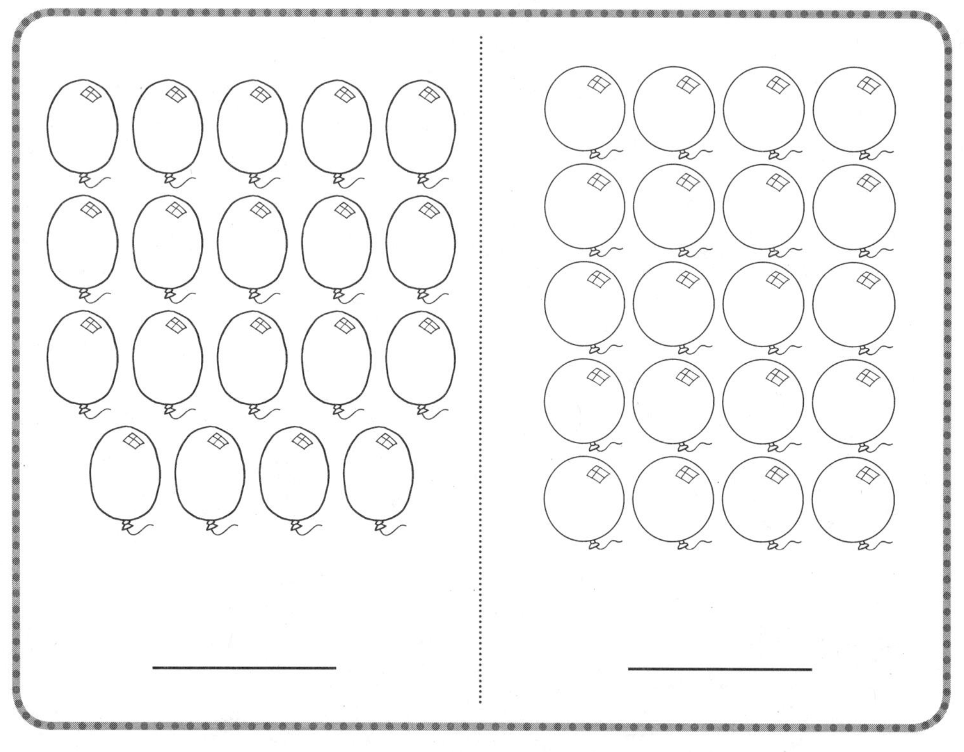

Name: ______________________________

Color each box when you complete that page.

① Introduction	② 1 to 20	③ 21 to 40	④ 41 to 60
⑤ 61 to 80	⑥ 81 to 100	⑦ 1 to 100	⑧ Review

Name: ______________________

Write the missing numbers.

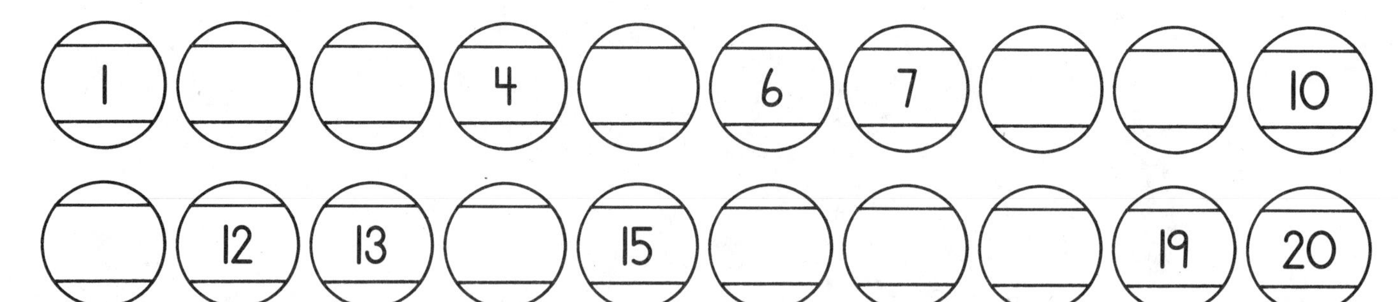

Say each number. Color that many cars.

17	
11	
20	
9	

Name: ____________________

Trace each number and number word.

30 30 thirty thirty

40 40 forty forty

Write the missing numbers.

How many? Circle the number.

28 35 39

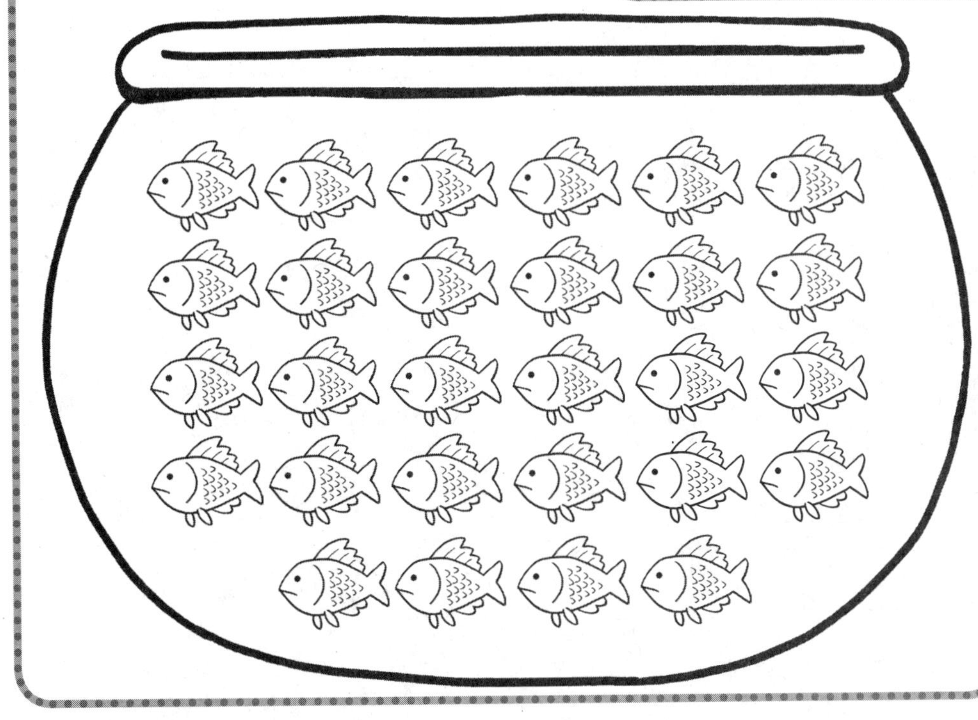

Name: ______________________________

Trace each number and number word.

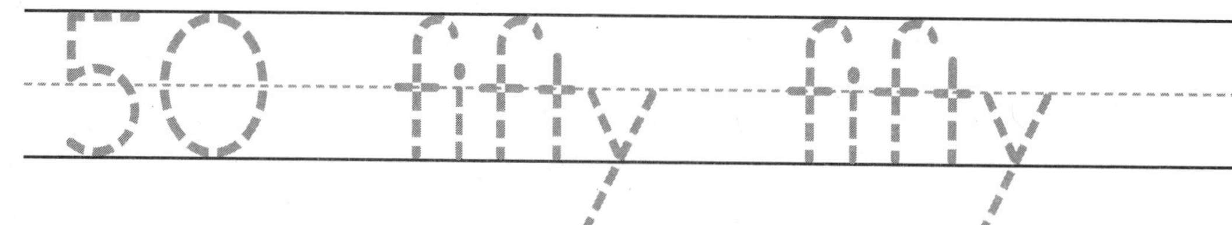

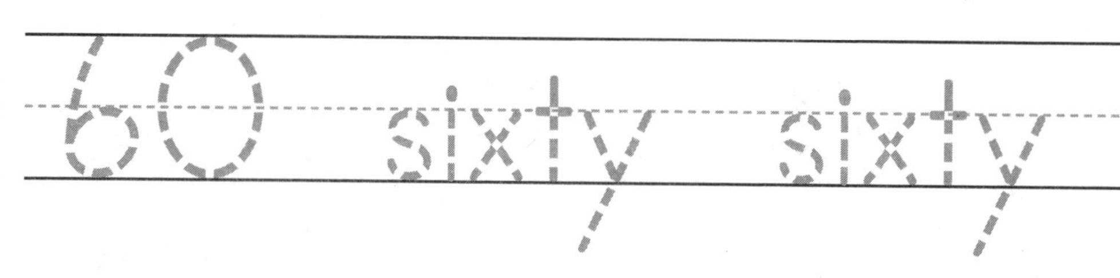

Write the missing numbers.

How many? Write the number.

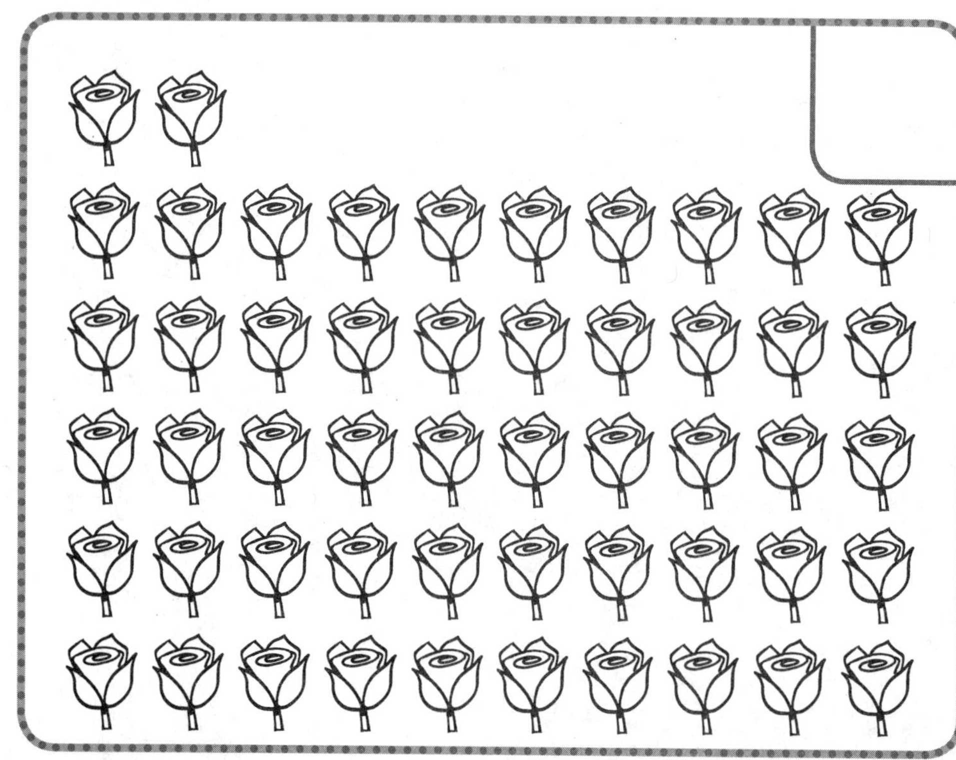

Name: ______________________

Trace each number and number word.

70 70 seventy

80 80 eighty

Write the missing numbers.

How many chocolate chips? Write the number.

Name: ____________________

NUMBERS 8
81 to 100
6

Trace each number and number word.

90 90 ninety

100 100 one hundred

Write the missing numbers.

How many? Circle the number.

79 89 98

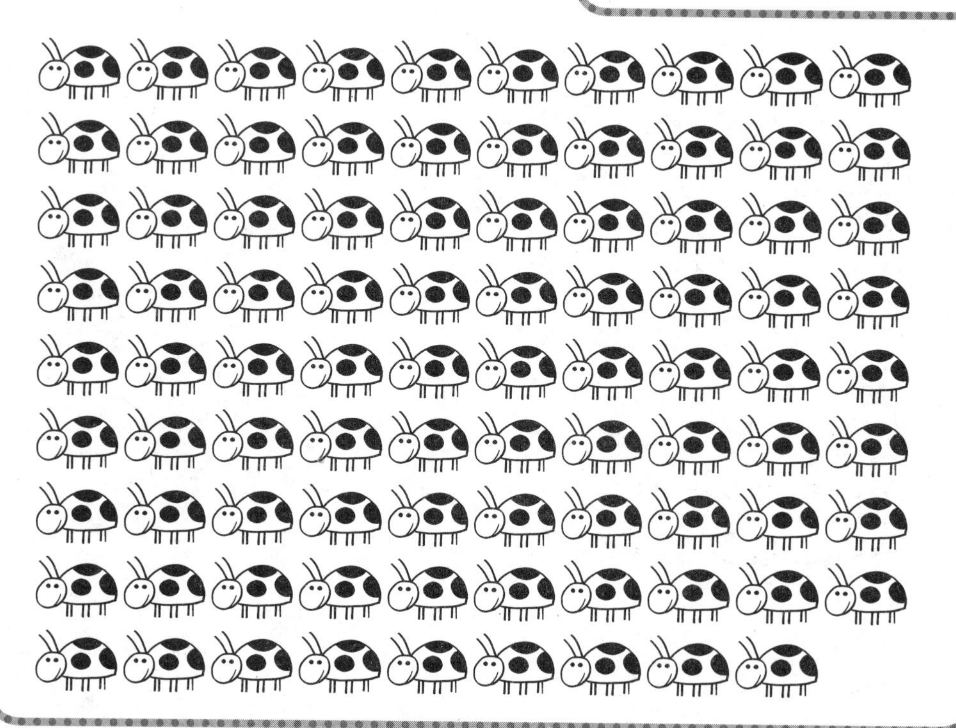

NUMBERS 8
1 to 100
7

Name: ____________________

Say each number!

Write the missing numbers.

1				5	6				10
	12	13					18		
21				25		27			
		33	34						40
41					46			49	
	52		54			57	58		
61				65		67			70
	72			75					80
		83			86		88	89	
91	92			95					100

Name: ______________________________

Write the missing numbers.

Which number is more? Color that butterfly.

How many? Write the number.

Name: ______________________________

Color each box when you complete that page.

1 Introduction	2 20, 30, 40	3 50, 60, 70	4 80, 90, 100
5 Sequence & Compare	6 Compare & Add	7 Count & Color	8 Review

Name: ______________________________

Match each number to its set.

40 • •

20 • •

30 • •

How many? Write the number.

NUMBERS 9
50, 60, 70
3

Name: ______________________________

Match each number to its set.

70 •

60 •

50 •

How many? Write the number.

Name: ______________________________

Match each number to its set.

100 •

80 •

90 •

How many? Write the number.

Name: ____________________

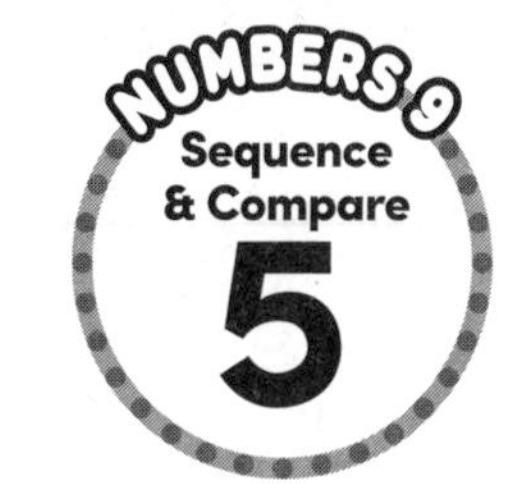

Count by tens. Write the missing numbers.

How many chocolate chips? Write the number.
Color the sundae that has more chocolate chips.

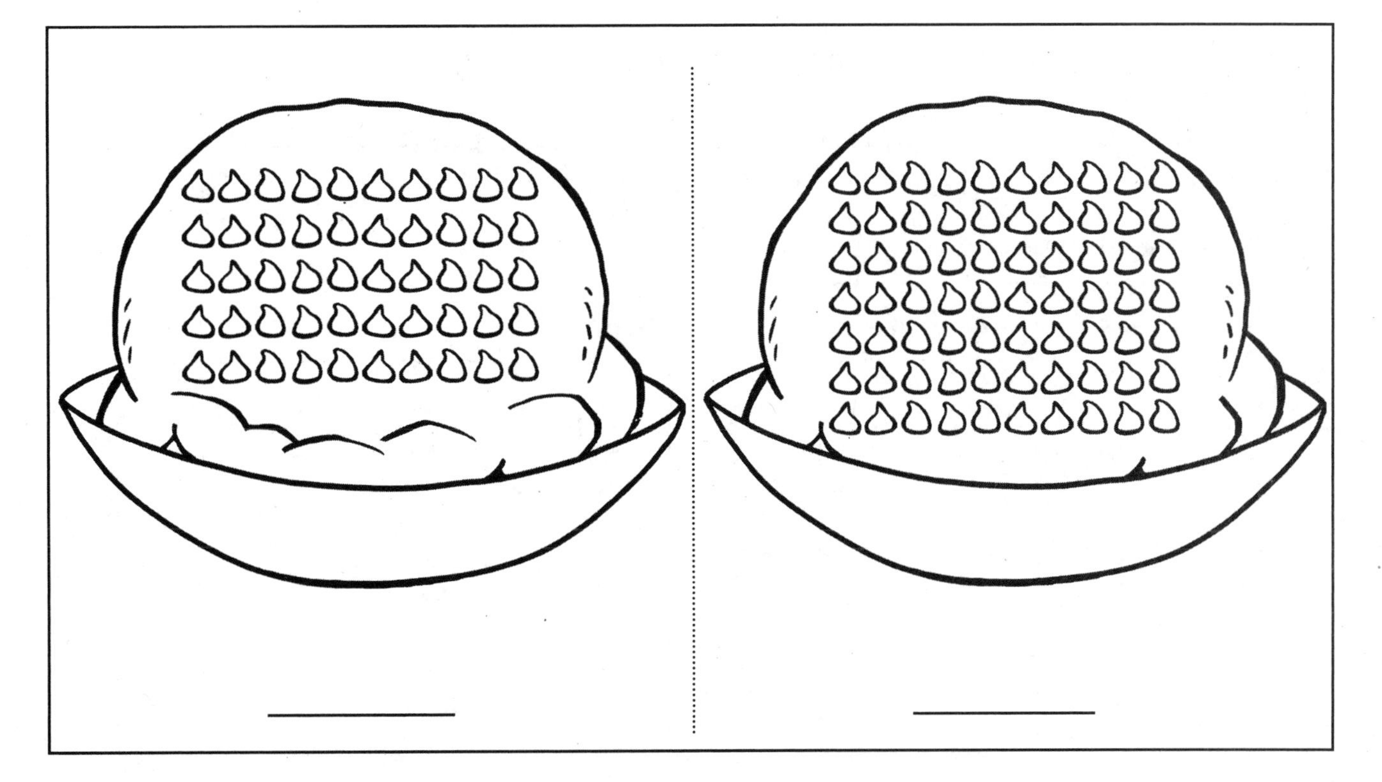

Name: ______________________________

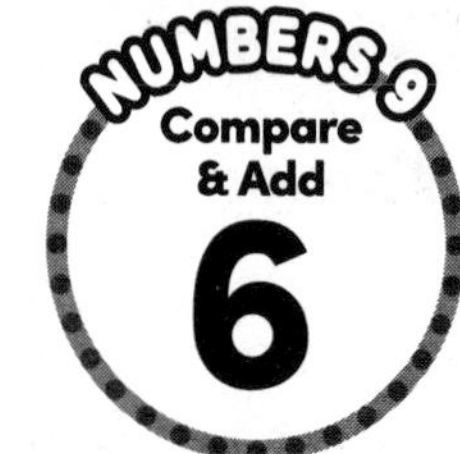

Count the bugs!

Circle the set that has fewer bugs.

Write ten more. Use the number line to help you. Then add.

START HERE.

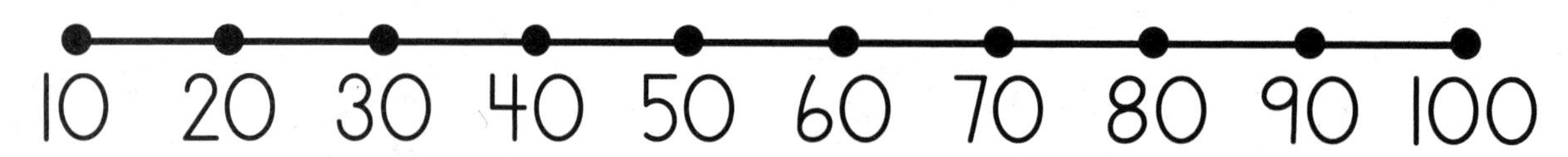

Number	10 more
50	
80	
20	

Add.

$50 + 10 =$ ______

$80 + 10 =$ ______

$20 + 10 =$ ______

Name: ____________________

Say each number. Color that many balls.

70	
100	

Say each number!

Use the code to color the cars.

thirty= red
forty= blue
fifty= yellow
sixty= green
eighty= orange
ninety= purple

90 50 40

30 80 60

NUMBERS 9
Review
8

Name: ______________________________

Trace each number. Then write the missing numbers.

10 20 ____ ____ 50

60 ____ ____ 90 ____

How many? Write the number. Circle the set that has more.

____

____

FUN PACK
Count & Color
1

Name: ______________________________

Use the code to color the picture.

Color each box when you complete that page.

1 Count & Color	2 Number Words	3 Connect-the-Dots	4 Sequence & Color
5 Complete the Path	6 Connect-the-Dots	7 Color by Number	8 Count & Write

Name: ______________________________

Read the number word. Write that number on the ice cream.

Name: ________________________________

FUN PACK
Connect-the-Dots
3

What is it?

Connect the dots. Start at 1.
Hint: Look for the star.

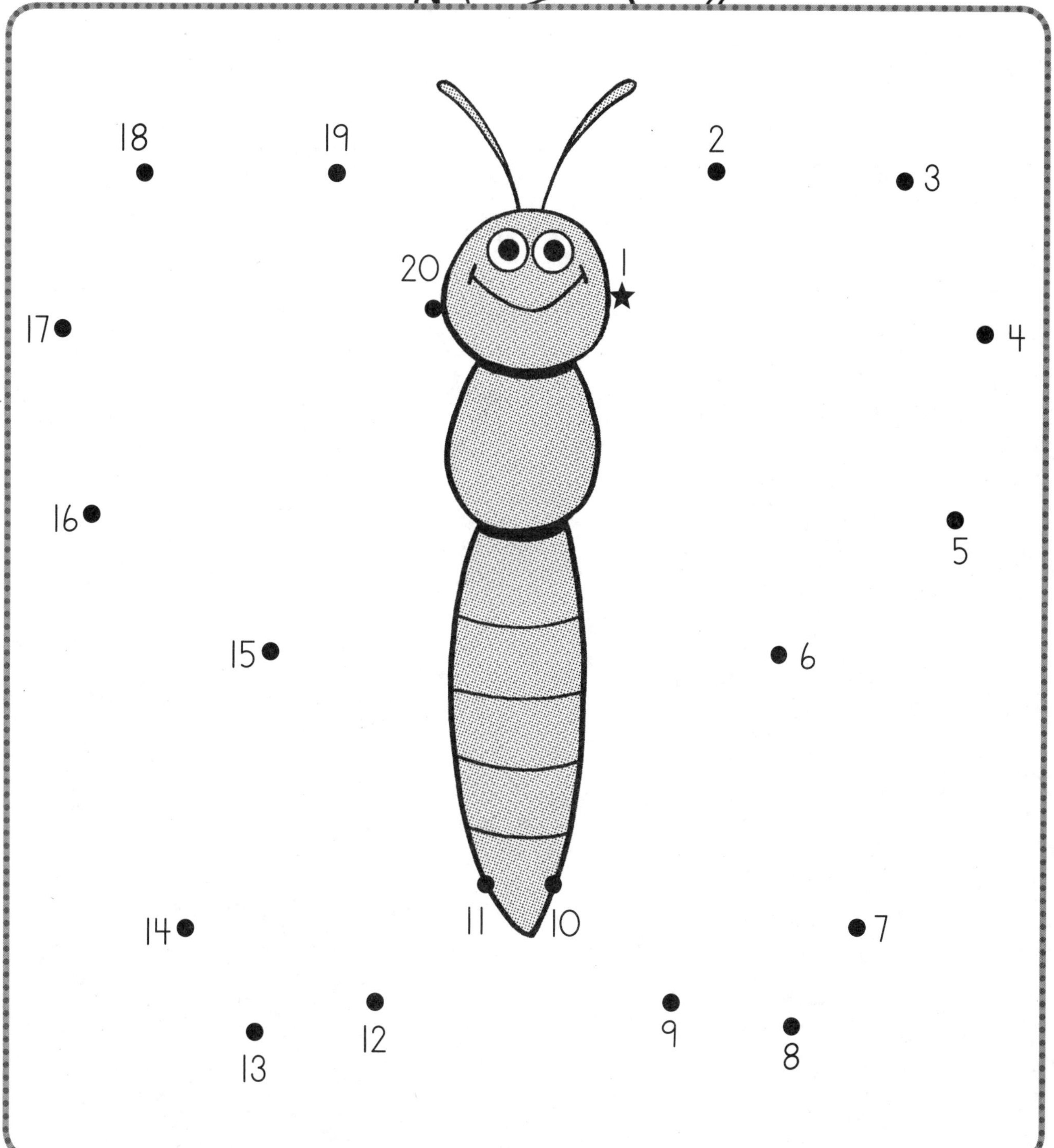

18 19 2 3 20 1 17 4 16 5 15 6 14 11 10 7 12 9 13 8

Name: ______________________________

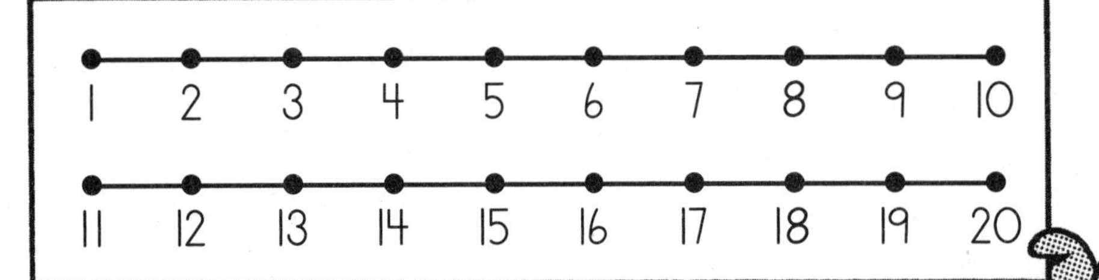

Write the missing numbers.
Then use the code to color the ants.

If the ant has a	7	10	16	19
Color the ant	red	yellow	orange	brown

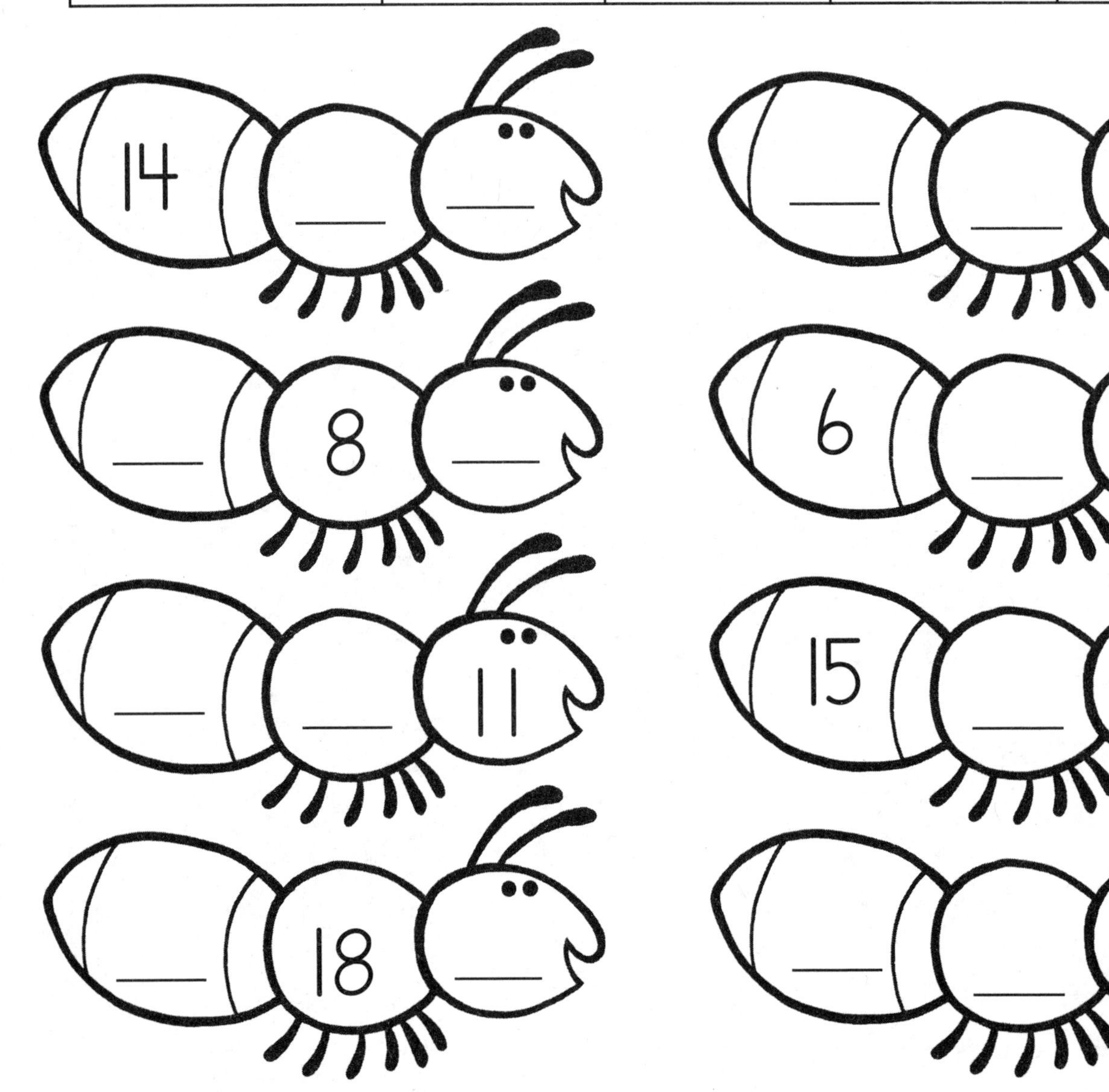

Name: ______________________________

Fill in the missing numbers from 21 to 40.

FUN PACK Connect-the-Dots 6

Name: ______________________________

Connect each set of dots. Start at 41 and 51.
Hint: Look for the stars.

Name: ______________________________

Use the code to color the picture.

If the number is from	Color the space
61 to 70	red
71 to 80	blue
81 to 90	yellow
91 to 100	orange

73 93 82 95 84 91 77 96 78 86 83 64 69 88 71 61 85 68 81 75 100 100

FUN PACK
Count & Write
8

Name: ______________________________

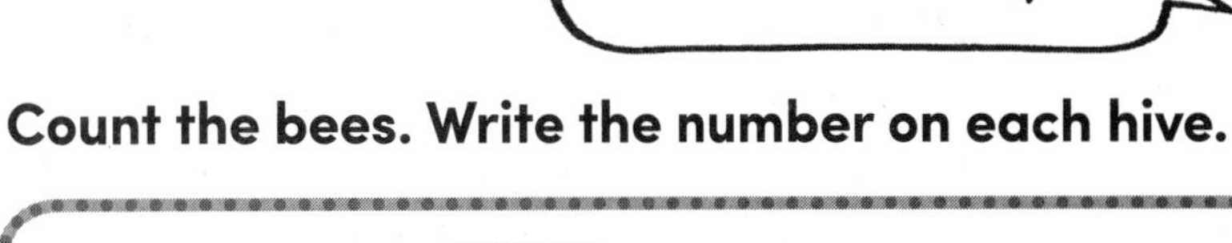

Count the bees. Write the number on each hive.

How many?

How many is **1** more?

How many?

How many is **10** more?

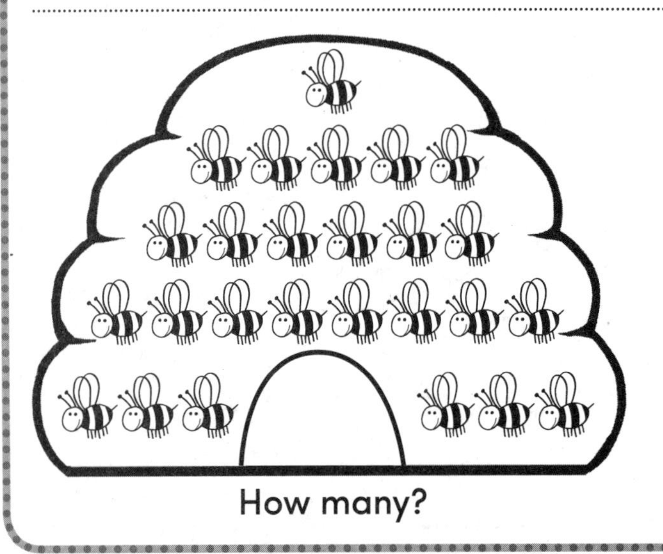

How many?

How many is **1** more?

Answer Key

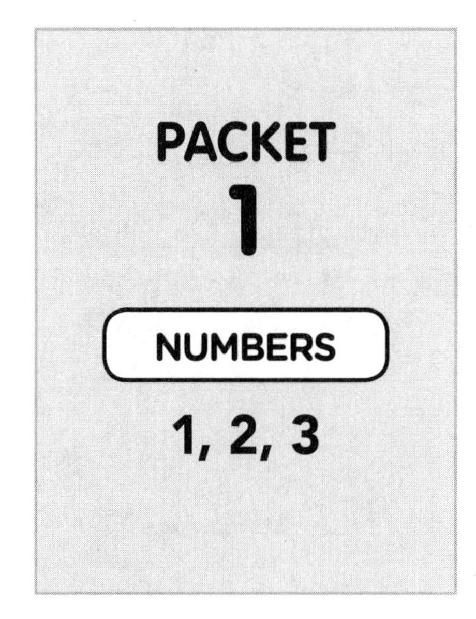

Name:

NUMBERS 1 Introduction 1

NUMBERS
1 2 3

Hi!

Trace each number.

1 1 1 1
2 2 2 2
3 3 3 3

Color each box when you complete that page.

1 Introduction	2 Count & Color	3 Count & Draw	4 Match & Write
5 Sequence & Compare	6 Compare & Add	7 Trace & Match	8 Review

7

Name:

NUMBERS 1 Count & Color 2

Say each number. Color that many balls.

2
1
3

Count the balls!

Use the code to color the cars.

1 = red 2 = blue 3 = yellow

RED YELLOW BLUE

8

Name:

NUMBERS 1 Count & Draw 3

Count the fish. Circle the number.

1 2 3
1 2 3

Draw 2 fish in the fish tank.

Count your fish!

9

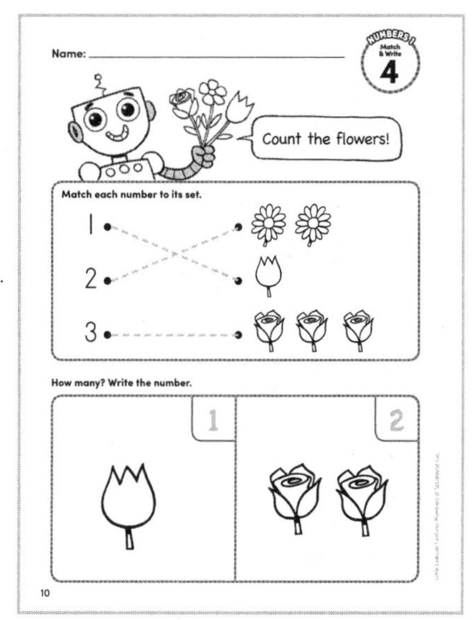

Name:

NUMBERS 1 Match & Write 4

Count the flowers!

Match each number to its set.

1
2
3

How many? Write the number.

1
2

10

Name:

NUMBERS 1 Sequence & Compare 5

Write the missing numbers.

1 2 3
1 2 3
1 2 3

Say the numbers!

YUM ICE CREAM

Color each cone that has more scoops.

11

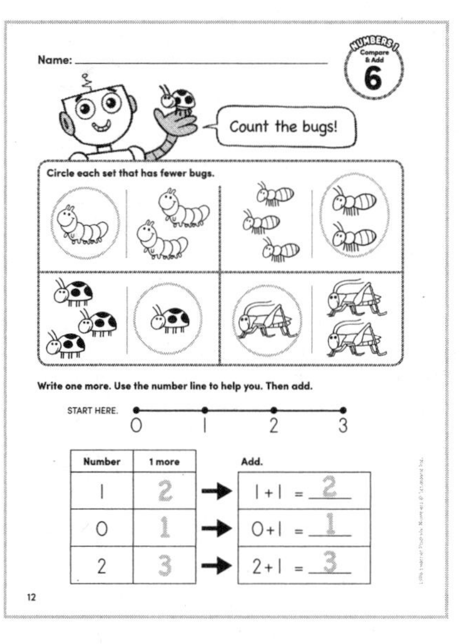

Name:

NUMBERS 1 Compare & Add 6

Count the bugs!

Circle each set that has fewer bugs.

Write one more. Use the number line to help you. Then add.

START HERE. 0 1 2 3

Number	1 more
1	2
0	1
2	3

Add.
1 + 1 = 2
0 + 1 = 1
2 + 1 = 3

12

Name:

NUMBERS 1 Trace & Match 7

Trace each number word.

1 one one
2 two two
3 three three

Read each word!

Match each number to its name.

2 • • one
3 • • two
1 • • three

13

Name:

NUMBERS 1 Review 8

Trace each number and word. Draw a set of balls in the box.

1 1 one
2 2 two
3 3 three

How many? Write the number. Circle the set that has more.

2 3
3 1

Great work! Bye!

14

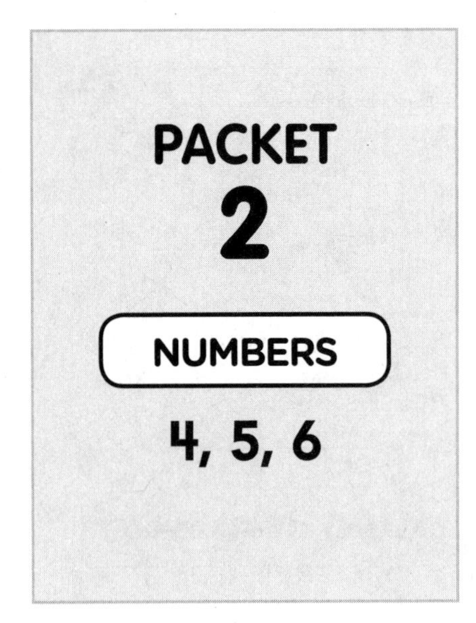
PACKET
2
NUMBERS
4, 5, 6

Name:
NUMBERS 2
Introduction
1
NUMBERS
4 5 6
Hi!
Trace each number.
Color each box when you complete that page.
1 Introduction
2 Count & Color
3 Count & Draw
4 Match & Write
5 Sequence & Compare
6 Compare & Add
7 Trace & Match
8 Review
15

Name:
Count & Color
2
Say each number. Color that many toys.
Count the toys!
Use the code to color the cars.
4= red
5= blue
6= yellow
BLUE
YELLOW
RED
16

Name:
Count & Draw
3
Count the fish. Circle the number.
4 5 6
4 5 6
Draw 4 fish in the fish tank.
Count your fish!
17

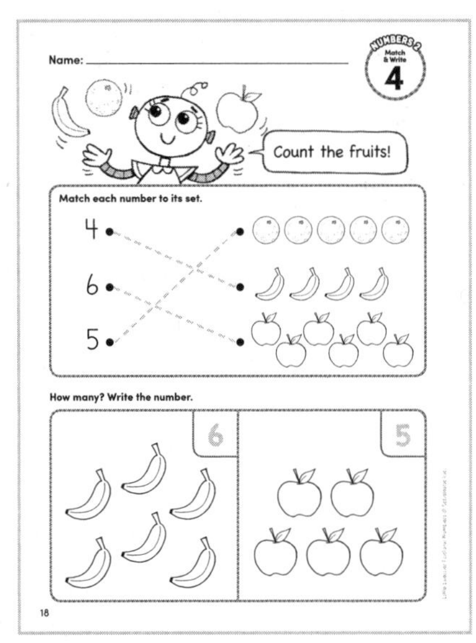
Name:
Match & Write
4
Count the fruits!
Match each number to its set.
How many? Write the number.
18

Name:
Sequence & Compare
5
Write the missing numbers.
Say the numbers!
YUM ICE CREAM
Color each sundae that has more chocolate chips.
19

Name:
Compare & Add
6
Count the bugs!
Circle each set that has fewer bugs.
Write one more. Use the number line to help you. Then add.
START HERE.
0 1 2 3 4 5 6
Number
1 more
Add.
4 + 1 =
3 + 1 =
5 + 1 =
20

Name:
Trace & Match
7
Trace each number word.
Read each word!
4 four four
5 five five
6 six six six
Match each number to its name.
five
four
six
21

Name:
Review
8
Trace each number and word. Draw a set of balls in the box.
4 four
5 five
6 six
How many? Write the number. Circle the set that has more.
Great work! Bye!
22

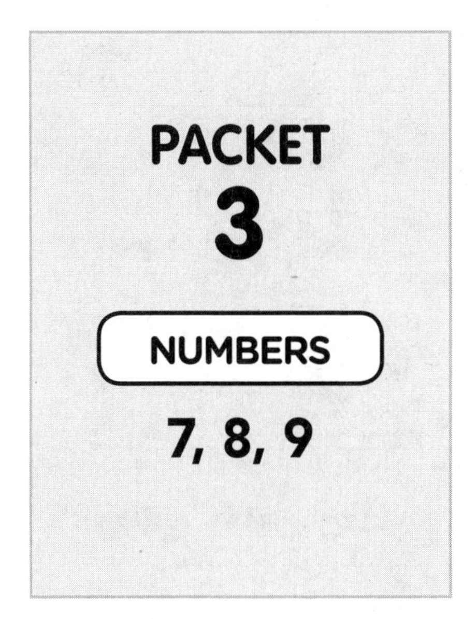
PACKET
3
NUMBERS
7, 8, 9

Name:
NUMBERS 3
Introduction
1
NUMBERS
7 8 9
Hi!
Trace each number.
7
8
9
Color each box when you complete that page.
1 Introduction
2 Count & Color
3 Count & Draw
4 Match & Write
5 Sequence & Compare
6 Compare & Add
7 Trace & Match
8 Review
23

Name:
Count & Color
2
Say each number. Color that many balls.
7
8
9
Count the balls!
Use the code to color the cars.
7=
red
8=
blue
9=
yellow
RED
YELLOW
BLUE
24

Name:
Count & Draw
3
Count the fish. Circle the number.
7 8 9
7 8 9
Draw 8 fish in the fish tank.
Count your fish!
25

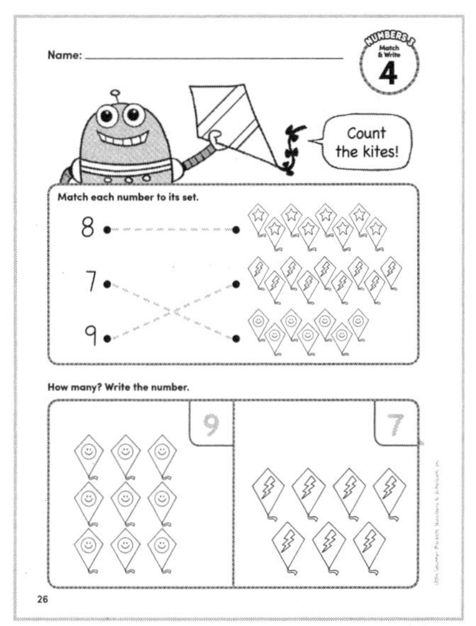
Name:
Match & Write
4
Count the kites!
Match each number to its set.
8
7
9
How many? Write the number.
9
7
26

Name:
Sequence & Compare
5
Write the missing numbers.
7 8 9
7 8 9
7 8 9
Say the numbers!
YUM ICE CREAM
Color each sundae that has more chocolate chips.
27

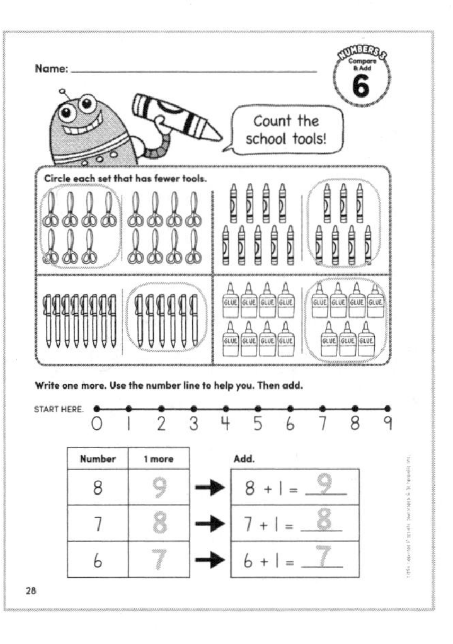
Name:
Compare & Add
6
Count the school tools!
Circle each set that has fewer tools.
GLUE
Write one more. Use the number line to help you. Then add.
START HERE
0 1 2 3 4 5 6 7 8 9
Number
1 more
Add.
8 9 8 + 1 = 9
7 8 7 + 1 = 8
6 7 6 + 1 = 7
28

Name:
Trace & Match
7
Trace each number word.
7 seven seven
8 eight eight
9 nine nine
Read each word!
Match each number to its name.
8
9
7
nine
seven
eight
29

Name:
Review
8
Trace each number and word. Draw a set of balls in the box.
7 7 seven
8 8 eight
9 9 nine
How many? Write the number. Circle the set that has more.
Great work! Bye!
9
8
30

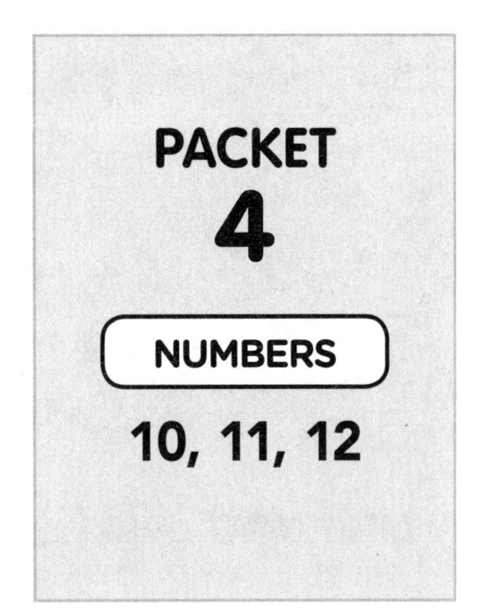

PACKET
4
NUMBERS
10, 11, 12

Name:
Introduction
1
NUMBERS
10 11 12
Hi!
Trace each number.
10 10 10
11 11 11
12 12 12
Color each box when you complete that page.
1 Introduction
2 Count & Color
3 Count & Draw
4 Match & Write
5 Sequence & Compare
6 Compare & Add
7 Trace & Match
8 Review
31

Name:
2
Say each number. Color that many toys.
11
10
12
Count the toys!
Use the code to color the cars.
10 = red
11 = blue
12 = yellow
BLUE
RED
YELLOW
32

Name:
3
Count the fish. Circle the number.
10 11 12
10 11 12
Draw 11 fish in the fish tank.
Count your fish!
33

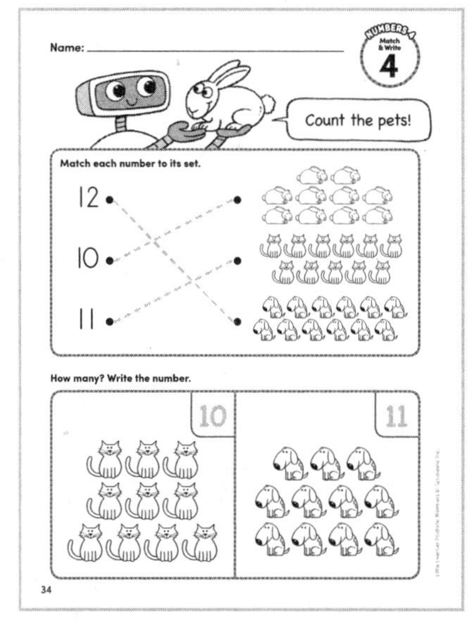

Name:
4
Count the pets!
Match each number to its set.
12
10
11
How many? Write the number.
10
11
34

Name:
5
Write the missing numbers.
10 11 12
10 11 12
10 11 12
Say the numbers!
YUM ICE CREAM
Color each sundae that has more chocolate chips.
35

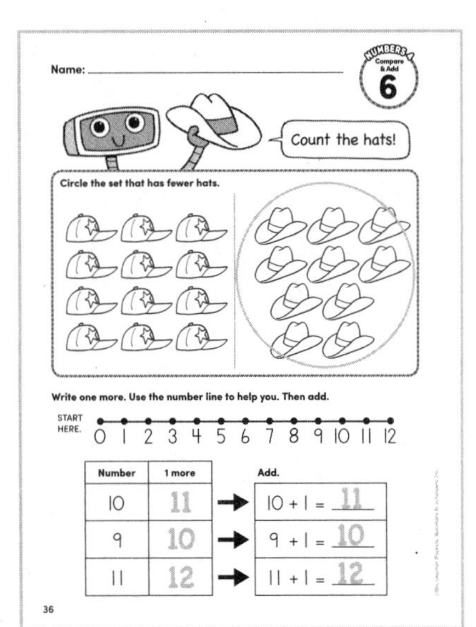

Name:
6
Count the hats!
Circle the set that has fewer hats.
Write one more. Use the number line to help you. Then add.
START HERE.
0 1 2 3 4 5 6 7 8 9 10 11 12
Number
1 more
Add.
10
11
10 + 1 = 11
9
10
9 + 1 = 10
11
12
11 + 1 = 12
36

Name:
7
Trace each number word.
10 ten ten ten
11 eleven eleven
12 twelve twelve
Read each word!
Match each number to its name.
12
ten
10
twelve
11
eleven
37

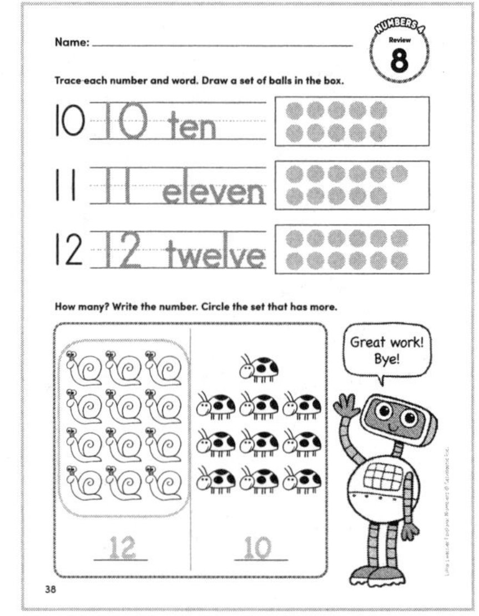

Name:
8
Trace each number and word. Draw a set of balls in the box.
10 10 ten
11 11 eleven
12 12 twelve
How many? Write the number. Circle the set that has more.
Great work! Bye!
12
10
38

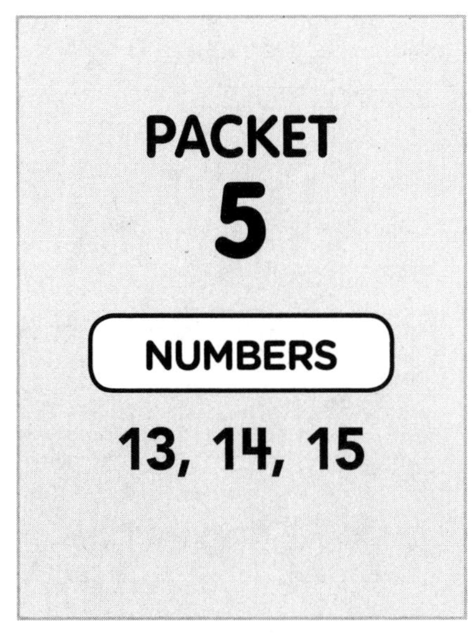

Name:

NUMBERS
13 14 15

Hi!

Trace each number.

13 13 13
14 14 14
15 15 15

Color each box when you complete that page.

1 Introduction	2 Count & Color	3 Count & Draw	4 Match & Write
5 Sequence & Compare	6 Compare & Add	7 Trace & Match	8 Review

39

Name:

Say each number. Color that many balls.

13
14
15

Count the balls!

Use the code to color the cars.

13 = red 14 = blue 15 = yellow

BLUE YELLOW RED

40

Name:

Count the fish. Circle the number.

13 14 15
13 14 15

Draw 14 fish in the fish tank.

Count your fish!

41

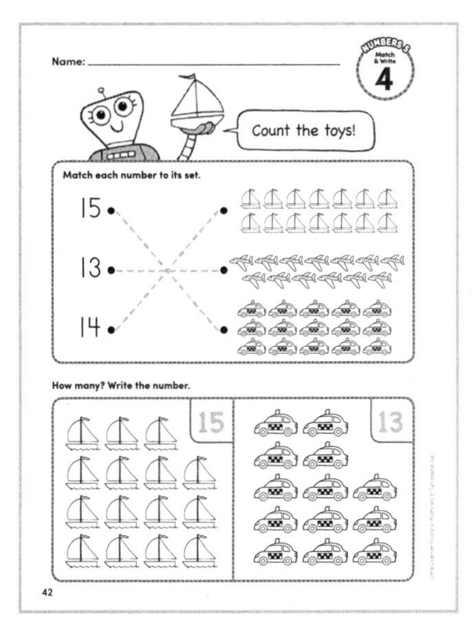
Name:

Count the toys!

Match each number to its set.

15
13
14

How many? Write the number.

15
13

42

Name:

Write the missing numbers.

13 14 15
13 14 15
13 14 15

Say the numbers!

YUM ICE CREAM

Color the sundae that has more chocolate chips.

43

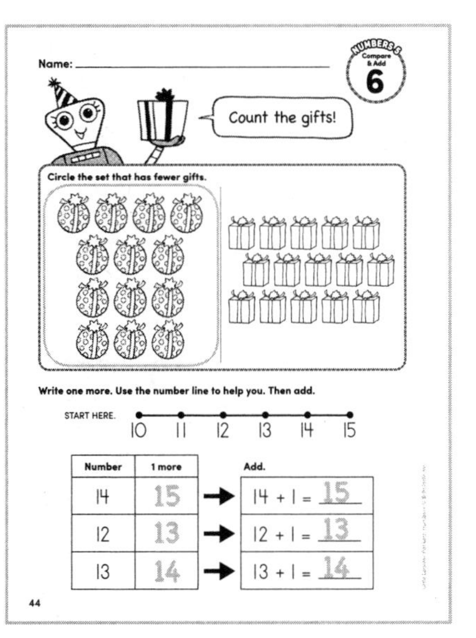
Name:

Count the gifts!

Circle the set that has fewer gifts.

Write one more. Use the number line to help you. Then add.

START HERE. 10 11 12 13 14 15

Number	1 more	Add.
14	15	14 + 1 = 15
12	13	12 + 1 = 13
13	14	13 + 1 = 14

44

Name:

Trace each number word.

13 thirteen
14 fourteen
15 fifteen

Read each word!

Match each number to its name.

13 thirteen
14 fifteen
15 fourteen

45

Name:

Trace each number and word. Draw a set of balls in the box.

13 13 thirteen
14 14 fourteen
15 15 fifteen

How many? Write the number. Circle the set that has more.

14 15

Great work! Bye!

46

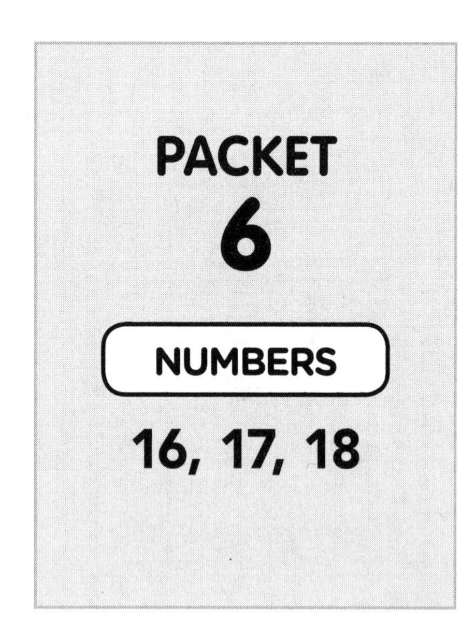
PACKET
6
NUMBERS
16, 17, 18

Name:
Introduction
1
NUMBERS
16 17 18
Hi!
Trace each number.
16 16 16 16
17 17 17 17
18 18 18 18
Color each box when you complete that page.
1 Introduction
2 Count & Color
3 Count & Draw
4 Match & Write
5 Sequence & Compare
6 Compare & Add
7 Trace & Match
8 Review
47

Name:
Count & Color
2
Say each number. Color that many toys.
18
16
17
Count the toys!
Use the code to color the cars.
16 = red
17 = blue
18 = yellow
YELLOW
RED
BLUE
48

Name:
Count & Draw
3
Count the fish. Circle the number.
16 17 18
16 17 18
Draw 17 fish in the fish tank.
Count your fish!
49

Name:
Match & Write
4
Count the foods!
Match each number to its set.
16
17
18
How many? Write the number.
14
16
50

Name:
Sequence & Compare
5
Write the missing numbers.
16 17 18
16 17 18
16 17 18
Say the numbers!
YUM ICE CREAM
Color the sundae that has more chocolate chips.
51

Name:
Compare & Add
6
Count the buttons!
Circle the set that has fewer buttons.
Write one more. Use the number line to help you. Then add.
START HERE.
10 11 12 13 14 15 16 17 18
Number
1 more
Add.
16 17 16 + 1 = 17
15 16 15 + 1 = 16
17 18 17 + 1 = 18
52

Name:
Trace & Match
7
Trace each number word.
16 sixteen
17 seventeen
18 eighteen
Read each word!
Match each number to its name.
18
16
17
sixteen
eighteen
seventeen
53

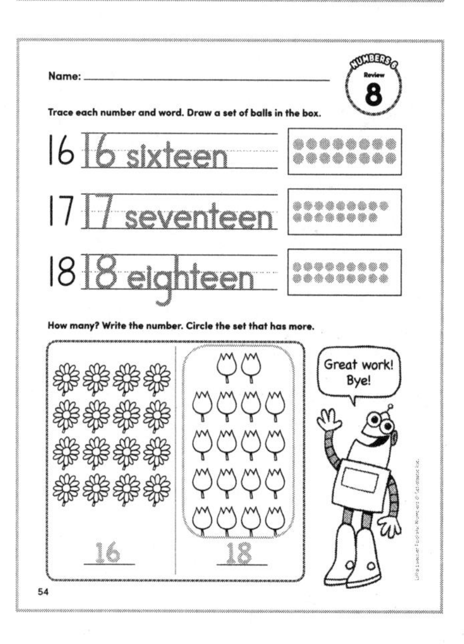
Name:
Review
8
Trace each number and word. Draw a set of balls in the box.
16 16 sixteen
17 17 seventeen
18 18 eighteen
How many? Write the number. Circle the set that has more.
Great work! Bye!
16
18
54

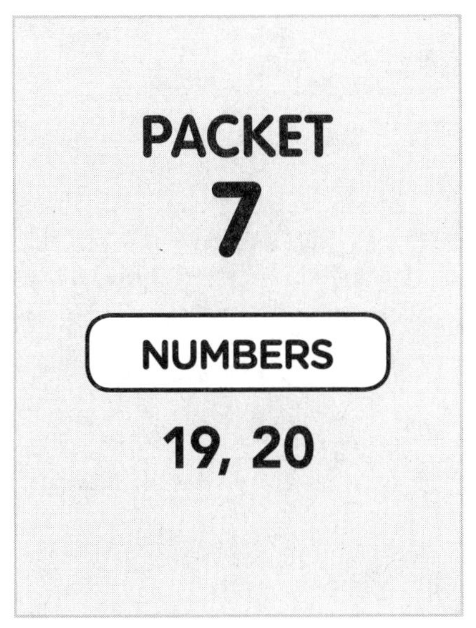
PACKET
7
NUMBERS
19, 20

Name:
NUMBERS 7 Introduction 1
NUMBERS
19 20
Hi!
Trace each number.
19 19 19
20 20 20
Color each box when you complete that page.
1 Introduction
2 Count & Color
3 Count & Draw
4 Match & Write
5 Sequence & Compare
6 Compare & Add
7 Trace & Match
8 Review
55

Name:
NUMBERS 7 Count & Color 2
Say each number. Color that many animals.
Count the animals!
20
19
Use the code to color the cars.
19 = blue 20 = yellow
BLUE
BLUE
YELLOW
56

Name:
NUMBERS 7 Count & Draw 3
Count the fish. Circle the number.
18 19 20
Draw 19 fish in the fish tank.
Count your fish!
57

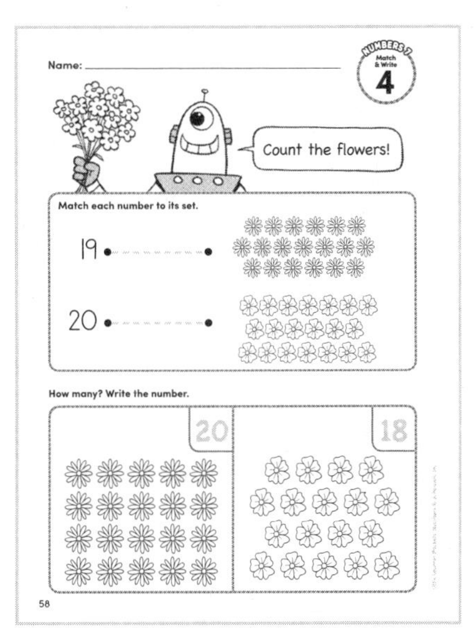
Name:
NUMBERS 7 Match & Write 4
Count the flowers!
Match each number to its set.
19
20
How many? Write the number.
20
18
58

Name:
NUMBERS 7 Sequence & Compare 5
Write the missing numbers.
18 19 20
18 19 20
18 19 20
Say the numbers!
YUM ICE CREAM
Color the sundae that has more chocolate chips.
59

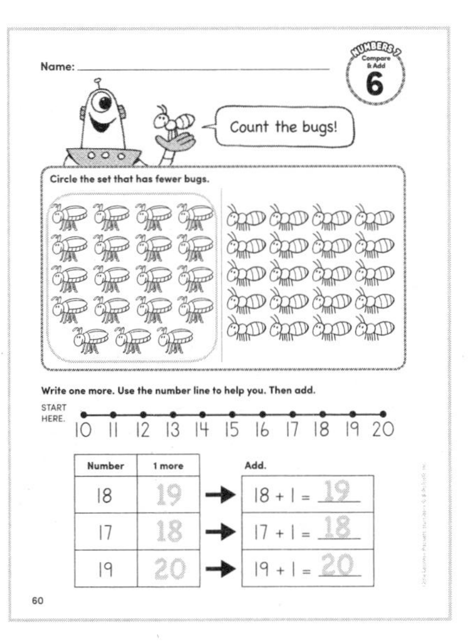
Name:
NUMBERS 7 Compare & Add 6
Count the bugs!
Circle the set that has fewer bugs.
Write one more. Use the number line to help you. Then add.
START HERE.
10 11 12 13 14 15 16 17 18 19 20
Number
1 more
Add.
18 19 18 + 1 = 19
17 18 17 + 1 = 18
19 20 19 + 1 = 20
60

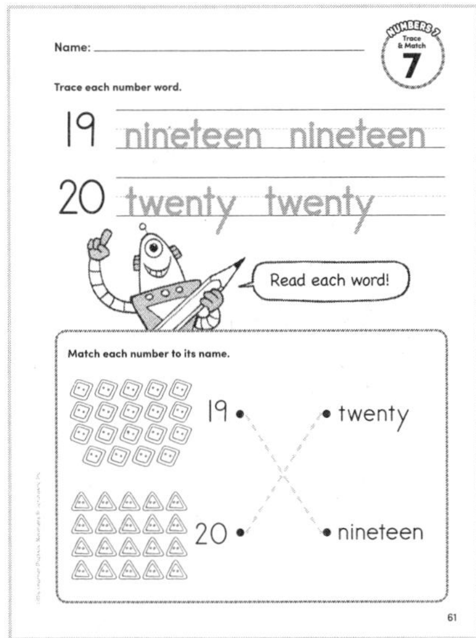
Name:
NUMBERS 7 Trace & Match 7
Trace each number word.
19 nineteen nineteen
20 twenty twenty
Read each word!
Match each number to its name.
19
twenty
20
nineteen
61

Name:
NUMBERS 7 Review 8
Trace each number and word. Draw a set of balls in the box.
19 19 nineteen
20 20 twenty
How many? Write the number. Circle the set that has fewer.
Great work! Bye!
19
20
62

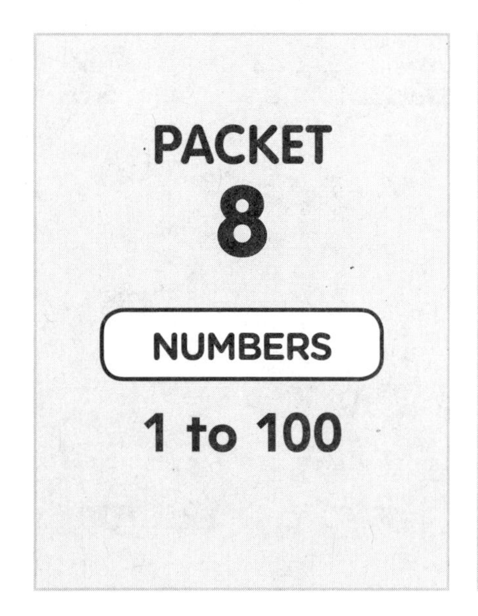
PACKET
8
NUMBERS
1 to 100

Name:
NUMBERS 8
Introduction
1
NUMBERS
1 to 100
Hi!
Trace each number and the number word.
100 100
one hundred
Color each box when you complete that page.
1 Introduction
2 1 to 20
3 21 to 40
4 41 to 60
5 61 to 80
6 81 to 100
7 1 to 100
8 Review
63

Name:
NUMBERS 8
1 to 20
2
Write the missing numbers.
Count the cars!
Say each number. Color that many cars.
17
11
20
9
64

Name:
NUMBERS 8
21 to 40
3
Trace each number and number word.
30 30 thirty thirty
40 40 forty forty
Write the missing numbers.
How many? Circle the number.
28 35 39
Say each number!
65

Name:
NUMBERS 8
41 to 60
4
Trace each number and number word.
50 50 fifty fifty
60 60 sixty sixty
Write the missing numbers.
How many? Write the number.
52
Say the number!
66

Name:
NUMBERS 8
61 to 80
5
Trace each number and number word.
70 70 seventy
80 80 eighty
Write the missing numbers.
How many chocolate chips? Write the number.
66
Say the number!
67

Name:
NUMBERS 8
81 to 100
6
Trace each number and number word.
90 90 ninety
100 100 one hundred
Write the missing numbers.
How many? Circle the number.
79 89 98
Say each number!
68

Name:
NUMBERS 8
1 to 100
7
Say each number!
Write the missing numbers.
69

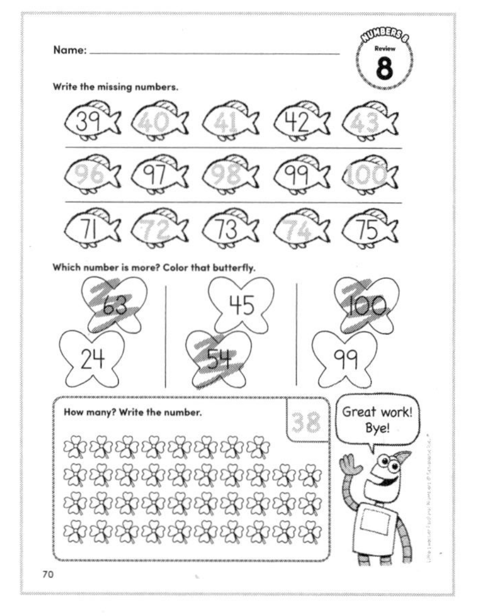
Name:
NUMBERS 8
Review
8
Write the missing numbers.
Which number is more? Color that butterfly.
63
24
45
54
100
99
How many? Write the number.
38
Great work! Bye!
70

PACKET
9
NUMBERS
10s to 100

Name:
Introduction
1
NUMBERS
10s to 100
Trace each number.
20 30 40
50 60 70
80 90 100
Hi!
Color each box when you complete that page.
1 Introduction
2 20, 30, 40
3 50, 60, 70
4 80, 90, 100
5 Sequence & Compare
6 Compare & Add
7 Count & Color
8 Review
71

Name:
2
Match each number to its set.
40
20
30
Count the toys!
How many? Write the number.
30
72

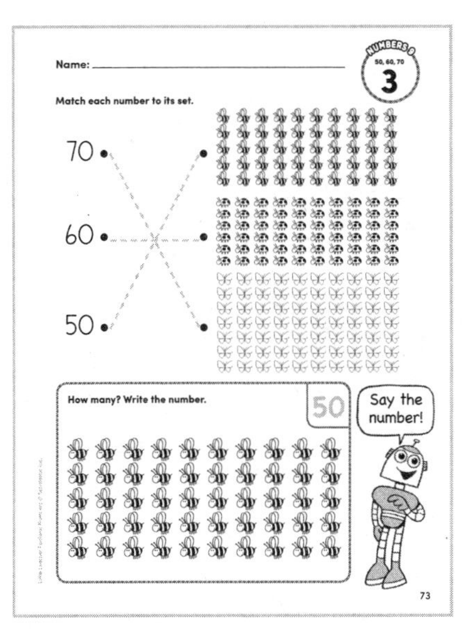
Name:
3
Match each number to its set.
70
60
50
How many? Write the number.
50
Say the number!
73

Name:
4
Match each number to its set.
100
80
90
Count the buttons!
How many? Write the number.
80
74

Name:
5
Count by tens. Write the missing numbers.
40 50 60
80 90 100
30 40 50
Say the numbers!
YUM ICE CREAM
How many chocolate chips? Write the number.
Color the sundae that has more chocolate chips.
50
70
75

Name:
6
Count the bugs!
Circle the set that has fewer bugs.
Write ten more. Use the number line to help you. Then add.
START HERE.
10 20 30 40 50 60 70 80 90 100
Number
10 more
Add.
50 60 50 + 10 = 60
80 90 80 + 10 = 90
20 30 20 + 10 = 30
76

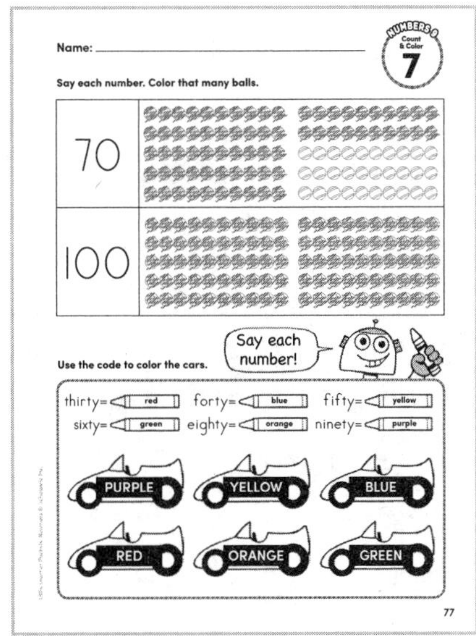
Name:
7
Say each number. Color that many balls.
70
100
Say each number!
Use the code to color the cars.
thirty= red
forty= blue
fifty= yellow
sixty= green
eighty= orange
ninety= purple
PURPLE
YELLOW
BLUE
RED
ORANGE
GREEN
77

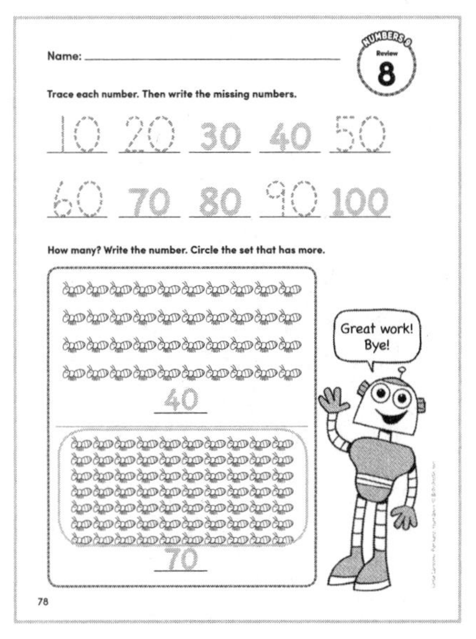
Name:
8
Trace each number. Then write the missing numbers.
10 20 30 40 50
60 70 80 90 100
How many? Write the number. Circle the set that has more.
40
70
Great work! Bye!
78

PACKET 10

10

NUMBERS

Fun Pack

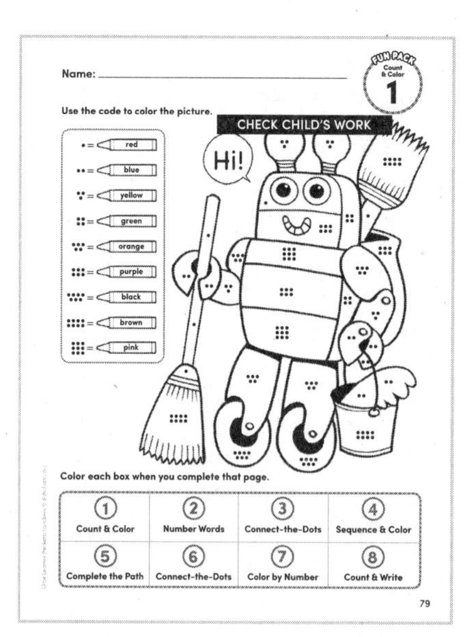

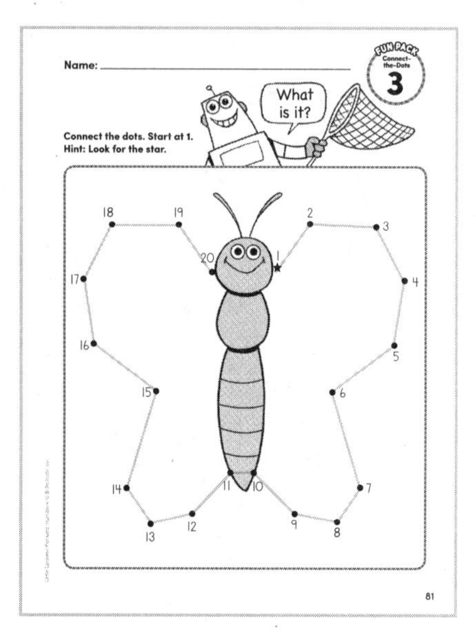

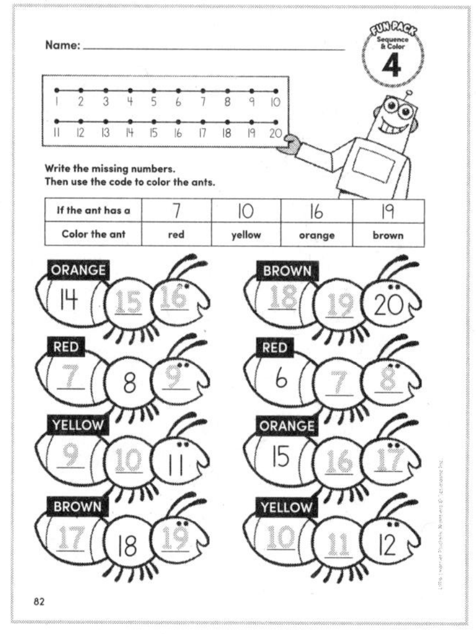

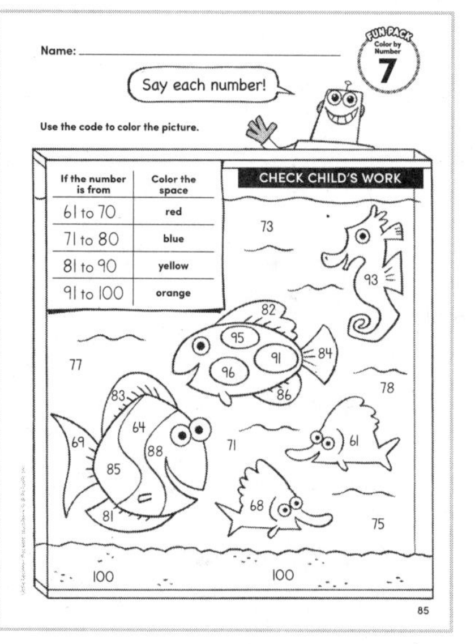